CW00602585

CraftFinder

the comprehensive guide to craft shops,
galleries and craft workers selling direct from their studios

incorporating " Craft Workshops in the English Countryside"

compiled and published by
THE WRITE ANGLE PRESS
in cooperation with
THE RURAL DEVELOPMENT COMMISSION

ISBN 0 9520737 4 9   ©**The Write Angle Press 1997**   **Price £9.95**

Country Cottage Crafts

FOR THE VERY BEST OF ART AND CRAFT FAIRS IN NORFOLK AND SUFFOLK. PLEASE TELEPHONE OR WRITE FOR A FREE BROCHURE AND LUXURY COACH EXCURSION DETAILS FROM NEIGHBOURING COUNTIES

- TOP QUALITY CRAFTSPEOPLE FROM ALL OVER BRITAIN •
- QUALITY REFRESHMENTS AT REASONABLE PRICES •
- FREE PARKING – CLOSER FOR THE DISABLED •
  FREE WHEEL CHAIR LOAN
- HAVE A DRINK AT "THE CRAFTERS ARMS" •
- CONTINUOUS CRAFT DEMONSTRATIONS •
- ALL EVENTS FROM 10 AM TO 5PM DAILY •
- FREE CHILDREN'S ENTERTAINMENT •

ADMISSION CHARGES
ADULTS £3 – SENIOR CITIZENS £2 – CHILDREN (5-15) £1.50

FOR FURTHER INFORMATION TELEPHONE
01986 788757

OR FAX
01986 788856

OR WRITE TO:

DAVID HICKS, TRUNCH HILL, DENTON, HARLESTON, NORFOLK IP20 0AE

CraftFinder

Contents   **Page**

research **Nina Massarik, Philip Stanbridge** • data entry **Maureen Tunningley**
cover **Hinton & Tunningley Design** • computer consultants **Associated Designs**
type and layout **The Write Angle Press** • film processing **Bureau 2000**
printing **Crewe Colour Printers** • distribution **Nice Bureau**

British Crafts
TRICIA LEIGH EXPOSITIONS

 presents events for 1997 and 1998

Kempton Exhibition Centre
Great Hall, Kempton Park Racecourse
Sunbury-on-Thames, Middlesex

13/14 September and 15/16 November 1997
and
14/15th March 1998 and 14/15 November 1998

Dollhouse and Miniature Fairs: 1st February and 6th December 1998

Location: Junction 1 of M3, Access from M25 Junction 12

Epsom Racecourse,
Rosebery Suites, The Grandstand,
Epsom Downs, Surrey

18/19 October and 6/7 December 1997
and
25th/26th April, 17/18 October and 5th/6th December 1998

Dollhouse and Miniature Fairs: 1st March and 6th September 1998

Location: M25 Junction 9,
Choice of 3 Rail Stations (Tattenham Corner nearest)

Lyndhurst, New Forest
Dollhouse and Miniature Fairs: 19th April and 18th October 1998

All events sign posted
All craft fairs open 10am to 5pm ...... both days
Ample free parking and refreshments at all events
Admission to Kempton Park and Epsom Racecourses
Adults £2.00, Senior Citizens £1.50 and Children Free (under 16)

Tricia Leigh Expositions
37 Parlaunt Road, Langley, Slough, Berkshire SL3 8BD
Enquireies Tel/Fax: 01753 545384

INTRODUCTION

Welcome to this, the first edition of CraftFinder, the comprehensive guide to finding and buying crafts in the UK. This book catalogues nearly 1000 places throughout England, Northern Ireland, Wales and Scotland that stock, make and sell craft work.

This work represents a fusion of two books. In the early course of researching a volume that was confined to craft shops and galleries we made contact with the Rural Development Commission who, for over thirty years, had published a directory of English Craft workshops. In discussion we learned that although their book was due for updating the Commission was not in a position to continue its production. CraftFinder as you now see it therefore represents the result of our work in cooperation with the Rural Developmnent Commission. It combines a list of craftworkers that sell direct from their premises with one of craft shops and galleries. While the list is as one, broken up only by geographic region, outlets of either type are easily identified as is explained on page 9.

The RDC publication extended to large scale crafts such as thatching, dry stone walling, forge work, large scale sad-dlery, stone masonry, craft brick work, carpentry and the like. Such crafts are not included in this volume but are to follow in a guide to rural trades that we expect to publish next year – the crafts found herein are, to press for a definition, those that might be readily picked up and taken away in the average car - albeit an estate version.

We have tried to make these listings as comprehensive as possible. There is no charge for inclusion and provided an outlet keeps regular hours that it is available to the public and has completed and signed a request for information they are entitled to an entry. We have not vetted any outlet or craft worker for quality of display or workmanship, believing such judgements are a matter between the reader and their purse or wallet.

We will be publishing a second edition of this book next year so please let us know of recommendation for improvement or of outlets that you would like to see added to our listings.

The Write Angle Press would like to thank the various UK craft guilds and associations as well as the Regional and National Tourist Boards for their help in the preparation of this book.

THE NATIONAL CRAFTS FAIR

National House
28 Grosvenor Road
Richmond, Surrey
TW10 6PB
Tel: 0181 940 4608
Fax: 0181 891 0115
Mobile 0370 626 424

HIGH QUALITY
CRAFT EVENTS IN 1997 AND 1998

Channel Islands

10th to 13th September 1997

9th to 12th September 1998

Beau Séjour Leisure Centre, St Peter Port, Guernsey

15th to 21st September 1997

14th to 20th September 1998

Fort Regent Leisure Centre, St. Helier, Jersey

Isle of Wight

25th and 26th October 1997

24th and 25th October 1998

St. Georges Park, Newport, I.O.W.

These Fairs are long established and proven events that provide variety, quality and good value in the crafts on display as well as a comfortable viewing environment for the visitor

*for further details contact **Anthony James** on*
0181 940 4608

In the first instance the book is arranged geographically by following the ten Tourist Board regions for England being Cumbria, East of England, Heart of England, Northumbria, North West, South East, Southern, West Country, Yorkshire and London plus the national tourist boards of Northern Ireland, Scotland and Wales. Full descriptions of the extent and character of each region will be found on pages 11 to 13.

Within these delineations the entries appear in alphabetical order according to the town (or nearest town in the case of small villages) where the outlet is located. When a town features more than one inclusion outlets are then arranged in alphabetical order according to the name of the shop. An index of all place names used in the book appears from page 198. An index of all the firms listed arranged alphabetically will also be found on page 193 as will an index arranged by craft type from page 189.

Those involved in the crafts are almost by definition individualists – determinedly so in many cases. As would-be individualists ourselves we have to applaud this in nearly all circumstances except in the case of cataloguing their activities in a book like this. So while what follows in this directory is an attempt at standardisation in conveying knowledge about craft outlets, please don't expect it to work in every case – for, as we've come to discover, different words mean very different things to different craft people.

As our introduction suggests, the outlets featured can be divided into two leagues – those with and those without craftworkers.

Firstly there are places where craft makers have based themselves to produce their work as well as selling direct to the public from these studios. These are easily identified in the listings as the phrase **Craftworkers on site:** will ap-pear within the entry and the number making work there will be shown. In addition, with these outlets the description of the goods sold will begin with the word **Crafts:** Often however a craft workshop or studio will broaden the appeal of their retail side by including work from other outside sources. Where, to the best of our knowledge, this is the case the description of the goods available will be expanded to read **Crafts/Stock:**. Remember such detail can and does change.

Sometimes, and it will be evident from the notes for such entries, there will be a group of craftworkers (sometimes as many as twenty) in one location working in different media from individual but grouped studios. Often in these locations there will be one shop selling all their products and a degree of cooperation over promotion and marketing. Many of these craft "centres" have sprung up in recent years often working from disused farm buildings in rural areas, old factories in urban ones or redundant churches in either. These have proved particularly successful as tourist attractions as a number of crafts can be seen at one time and the studios are often open for visitors to view work in progress. Scale has often allowed the establishment of other facilities such as tea rooms, picnic places, child play areas etc. which will be noted in the entries for such establishments.

Secondly there are craft shops and galleries where a straightforward retail operation takes place. Here the owner will buy or take on sale or return craft goods from outside craft makers and sell these works to the general public – simple enough and the same as most retail shops anywhere. These are readily picked out in the listings as no reference will be made to Craftworkers on Site and the goods available text will be preceeded by the one word **Stock:** Please be aware, as we've already said, this is just a rule of thumb as definition

jewellery

Textiles

Ceramics

ACCESSORIES

Knitwear

CRAFTS

Over 100 Scottish companies under one roof
–
**10 miles from Inverness in the historic
village of Beauly**

The Made in Scotland Shop
The Craft Centre,
Beauly, Inverness-shire IV4 7EH
Telephone 01463 782821

MADE IN
SCOTLAND

see full entry on page 137 of this book

is not easy even where proprietors are entirely cooperative with the regime of questions. Often they have wanted to suggest that our definitions are entirely arbitrary, which of course they are, or inappropriate, which in the context of this book we believe them not to be.

In nearly every case the information contained in this book has been based on a questionnaire completed by the proprietor of the businesses concerned. This information has been checked by way of returning a proof copy of the entry to the business to allow it to be checked and incorrect or changed data to be revised. Indeed, in many cases where items of information remained unclear or missing even after this stage direct contact by telephone was used to complete the picture. In short we have gone to considerable lengths to ensure that the entries contained in this book are accurate at the time of going to print. However we have accepted the information provided in good faith and as such are not able to accept responsibility for the veracity of statements made in the book or accept any responsibility for loss financial or otherwise that might result from error or omission in this work. We strongly suggest therefore that you confirm any information upon which you intend to rely with the outlet concerned.

We would particularly draw your attention to the matter of opening hours. For the most part we have excluded outlets that were unwilling or unable to give us specific hours. Where supplied with this information it is shown, so to the best of our knowledge the hours given are those that these businesses are open to the public. We have to say that we have noted a certain *flexibility* in the way crafts people tend to interpret opening and closing times. We must strongly suggest therefore that you telephone any outlet that you intend to visit before starting out – particularly if you have a lengthy journey to make. A similar point can be made about the crafts and stock available at each out-

let. We have listed these items as we understood them to be at the time of going to press but as such ranges are often an expression of the preferences of the proprietor concerned these can change. If you are after a specific item check its availability on the phone before leaving home.

It is important to note also that the price ranges provided are **very** approximate and are only intended to give an additional guide to the character of the outlet concerned.

Direct commissioning of craftworkers to make specific items is an accepted, indeed welcome practice. Notes about this will be found within many entries for craftworkers in the main listings but the list of craft guilds and societies on page 176 will also assist. If you have been unable to find the exact work of, say, weaving that you want, use this list to make contact with a local branch of the Guild of Spinners, Weavers and Dyers where the contact person given in the list should be able to put you in touch with a member of that guild able to accept your commission.

When visiting a craft worker in his or her studio do remember that it is their place of work (sometimes even their home) so please respect it and their working time. Don't be too demanding of their time or knowledge and keep your visit as short as is appropriate. If at all possible buy something.

There is in the UK a quite well developed network of craft fairs – indeed there are probably over 3000 such events held each year representing a wide variety of style and content. We publish another title, **The Craftworker's Year Book**, that carries a comprehensive diary of such events and while we have not sought to reproduce that here we have made a selection of those organisations who, in our opinion, put together the best organised events. By telephoning any of these you should be able to obtain a full list of their events.

As mentioned on page 9, this book is subdivided into thirteen sections following the English Regional Tourist Board boundaries together with the national ones of Northern Ireland, Wales and Scotland. A map appears over the page to illustrate how they are delineated.

Cumbria, the first of our regions, is situated in the most north westerly corner of England with Scotland and the sea as neighbours. It is a region popular with visitors because of its phenomenal physical beauty. When we think of Cumbria we think first of the Lake District and names such as Windermere, Thirlmere and Coniston which have been etched into our national psyche by Wordsworth and his fellow "Lakeland Poets". While the beauties of the Lake District National Park are certainly the main theme, the periphery marked by the mountains running out into the varied surrounding landscapes of coasts, hills and dales should not be something to pass by quickly.

East of England is at the opposite end of the country and is "opposite country" – for even its most precipitous slopes are never more than rolling with much of it flat old fenlands – now intensively farmed. The region comprises the counties of Norfolk, Suffolk and Essex – the flat ones with north sea coastlines – with Bedfordshire, Cambridgeshire and Hertfordshire stretching inland and being more rolling in nature. Toward the end of last year the Regional Board known as East Midlands was disbanded and Lincolnshire was added to those counties above to make up this newly named region. The Norfolk Broads, the bracing coastal resorts, picturesque villages and the lack of gradient for those on foot or two wheels have been traditional attractions for visitors. Improvement to transport to the area – in particular the completion of the M11 and its link to the M25 – has increased the popularity of the region for weekend breaks.

Heart of England, as the name suggests, is at the centre of the country and extends from the border with Wales in the west to the East of England counties in the east. The huge West Midlands conurbation around Birmingham sits in turn at the heart of the area. This region too was expanded with the demise of East Midlands region and thus has become the most populated of the regions. Our listings for this one are the most extensive and of all the regions it probably represents the greatest variety of craft outlets from the studios and

grouped workshops of the urban areas to the prestigious galleries of some of the rural areas.

Northumbria, England's most northerly region has perhaps been unfairly overlooked in the terms of popularity with tourists. To be kind the climate can often merit the euphemism "bracing". However, for the more discerning visitor it has a huge amount to offer. The many protected "Areas of outstanding natural beauty", the Northumberland National Park and spectacular coastlines dotted with their craggy castles offer brilliant walking experiences for the well clad. Glass working has been a tradition of the area and this survives both in the galleries and the studios of the region with The Glass Trail and The National Glass Centre based in Sunderland.

The counties of Cheshire, Lancashire and the high peak area of Derbyshire as well as the metropolitan areas of Merseyside and Manchester comprise the **North West** region. The region has a great industrial history connected with the textiles industry and the cotton industry in particular. The names of mill towns such as Oldham, Burnley or Wigan are synonymous with that 19th century growth as eventually they became with the vagaries of the decline that came with the lingering demise of "King Cotton". The string of seaside resorts that grew up at that time survived the decline with much of their Victorian gentility intact and Lytham, Morecambe, Southport and Blackpool are still important attractions for visitors to the area. The region has a good measure of scenic beauty – most notably in the rugged contours of the Peak District National Park. The craft scene is varied and rich also, with again a combination of urban workers sometimes operating from those old industrial buildings as well as one of the best known UK galleries – The Bluecoat Display Centre in the centre of Liverpool.

The **South East** region is made up simply of four counties. A good part of the north of the region is built up where it is adjacent to London and perhaps does not hold the fascination for the visitor that the Kent apple orchards in spring or the Wealden countryside of Sussex or Surrey can command. The region presents miles of coastline to the channel which are a primary visitor attraction with such resorts as Brighton, Hastings and Eastbourne seeming to retain their traditional pull. The region is rich in craft of all types - particularly to the west where traditional crafts such as trug making

The English Regional Tourist Board Boundaries

SCOTLAND

NORTHUMBERLAND
NORTHUMBRIA

TYNE AND WEAR

DURHAM
CUMBRIA
CUMBRIA
CLEVELAND

NORTH YORKSHIRE
YORKSHIRE

EAST
YORKSHIRE

LANCASHIRE
WEST
YORKSHIRE

GREATER
MANCHESTER
MERSEYSIDE
SOUTH
YORKSHIRE
**NORTH
WEST**
CHESHIRE
DERBYSHIRE
LINCOLNSHIRE
NOTTINGHAM-
SHIRE

STAFFORD-
SHIRE
NORFOLK

SHROPSHIRE
LEICESTERSHIRE
WEST
MIDLANDS

WALES
NORTHAMPTON-
SHIRE
CAMBRIDGE-
SHIRE
SUFFOLK
**HEART OF
ENGLAND**
WARWICK-
SHIRE
BEDFORD-
SHIRE
HEREFORD AND
WORCESTER
BUCK-
INGHAM-
SHIRE
EAST OF ENGLAND
GLOUCESTERSHIRE
HERTFORD-
SHIRE
OXFORD-
SHIRE
ESSEX
GREATER
LONDON
LONDON
NORTH
SOMERSET
BERKSHIRE
KENT
WILTSHIRE
SURREY
**SOUTH EAST
ENGLAND**
SOMERSET
SOUTHERN
HAMPSHIRE
WEST
SUSSEX
EAST
SUSSSEX
WEST COUNTRY
DORSET
DEVON

ISLE OF
WIGHT
CORNWALL

0 10 20 30 40 50
Miles

0   20   40   60   80
Kilometres

can be found in a number of places as well as a strong theme of contemporary furniture.

The **Southern** region is long and thin and, as the map shows, extends in a seemingly unlikely fashion from Oxford in the north to the south coast and the Isle of Wight as well as extending sharply west to include part of Dorset . The area has for many years been a region of relative prosperity which seems to have provided a climate for a thriving arts and crafts scene with a number of well known galleries across Hampshire and Dorset as well as a rich seam of craft producers. Shopping for crafts in the region is made even more pleasurable by the countryside being about as typically English as you can get with pretty villages and traditional market towns set out among gentle hills and open fields linked, if you keep off the 'A' roads, by winding country lanes. The coastline has its traditional resorts such as Bournemouth or Lyme Regis as well as 80 miles of protected coastline much of it managed by the National Trust. The region also includes the New Forest.

The **West Country** in terms of latitude is the longest of the regions stretching as it does from Lands End and the granity coastline of Cornwall in the west through the rolling red farmland of Devon to the prairie-like fields of Salisbury plain. There is a long tradition of crafts in the region. The clay deposits in Devon and Cornwall have supplied potteries for hundreds of years and come the 20th century St.Ives was the location for Bernard Leach to establish the new movement in studio pottery. There are still excellent potters throughout the region. The Somerset levels, a low lying area, has a similarly long tradition of growing willow for baskets and that continues.

Yorkshire has long been famed for the landscape of its dales and moors but of recent years this has taken on almost cult status through the success of various TV programmes such as "Last of the Summer Wine", "All Creatures Great and Small" and others. Like Northumbria, its neighbour, the climate can sometimes be severe here and wet but the properly kitted out walker can enjoy over 1000 square miles of protected areas of Natural Beauty, National Parks or Heritage Coastline. Traditional Victorian resorts such as Harrogate, Scarborough or Whitby retain their importance for their visitors. While most of the crafts are very well represented in the region ceramics, perhaps drawing on a tradition of the area, is particularly strong both in makers and galleries to display.

We will deal only briefly with **London** that covers the area that was, until abolition, the GLC administrative region. The primary attraction for the visitor is of course London as a capital city. A wide of variety of crafts are represented by galleries and makers with an emphasis on the non-traditional, not to say wacky.

Northern Ireland offers a similar combination of splendid countryside and "craic" to that found south of the border, with tourism and crafts both on the up. Areas of outstanding physical beauty and natural interest abound – the Mourne, Antrim and Sperrin mountains, Lough Neagh, Loug Erne, Strangford Lough, the Giant's Causeway and the northern coastline. Visitors usually try to take in the Bushmills Distillery, Belleek Pottery, Tyrone Crystal, The Irish Linen Centre and of course the attractions of the capital – Belfast.

Scotland is one administrative region for the purpose of the promotion of tourism and incorporates all its regions. Scotland attracts huge numbers of visitors over the summer months with an upward trend quite noticeable over recent years. The attractions are myriad with phenomenal scenery – even in wet weather – of lochs, mountains, glens and rocky coasts. The major towns boast many cultural attractions and Edinburgh Festival has long since reached international standing in the performing arts. Organisations such as Made in Scotland have directed considerable effort to developing craft business north of the border and, as the volume in that section shows, with some success. A wide variety of crafts can be seen although it has to be said that for the most part they are fairly traditional in their design and execution. The Scottish Islands (listings are separated to the back of the section for ease of use) also have a well developed craft network, largely orientated to the summer visitor.

All of **Wales** is also dealt with by one National Tourist Board. They too have made great progress in promoting their craft industry to a wider world market. The Wales Craft Council has been significant in that process. Wales has long attracted visitors with its variety of landscape - some of the highest mountains in the UK, leafy rural tracts with rivers running through them and hundreds of miles of coastline with both rocky and sandy shorelines to enjoy. The crafts cover again the full range and while there are a number of galleries specialising in contemporary and experimental work, the tendency is toward the traditional with much work having a defined Welsh identity

GOSSIPGATE GALLERY
The Butts
Alston
Cumbria CA9 3JU
Tel: 01434 381806
Contact: Mrs. Sonia Kempsey
Established: 1983
Open: 10am to 5pm daily including Sundays Easter-October. Winter weekends 10.30am to 4.30pm, weekdays variable; telephone for details. Closed January to mid-February.
Stock: A wide range of northern art and craft - pottery, sculpture, paintings and prints, wood, glass, jewellery, knitwear. Also cards, books of local interest, cassettes and CDs of Northumberland pipe music, Cumberland mustard and local jams and preserves.
Prices: 99p to £600. Amex, Visa, Access, Mastercard accepted.
Notes: Coffee shop and tea garden. Garden pots and herbs.

THE PENNINE POTTERY
Clargill Head House
Alston
Cumbria CA9 3NG
Tel: 01434 382157
Contact: Peter Lascelles
Established: 1992
Open: April 1st to January 1st 10am to 5.30pm. February to March weekends 10 am to sunset.
Craftworkers on site: 1
Crafts/Stock: Stoneware pottery mainly functional (mugs, jugs, ovenware etc.) Terracotta flower pots. Wooden items, hooky mats, watercolours, knitting and handweaving.
Prices: £2 to £100.
Note: Simple lunches, teas, snacks. Bread baked daily. Homemade soups, cakes etc. Pottery display in a part of our Gift Shop. 2.5 miles east of Alston on A686. Car park. Reasonable disabled access. Children welcome.

ADRIAN SANKEY GLASS
Rydal Road
Ambleside
Cumbria LA22 9AN
Tel: 015394 33039
Fax: 015394 31139
Contact: Adrian Sankey

Established: 1981
Open: 9am to 5.30pm
Craftworkers on site: 5
Crafts: Handmade lead crystal glass.
Prices: £3 to £300. Visa and Mastercard accepted.
Note: Demonstrations all day of glass making. Very distinctive restaurant.

DEXTERITY
Kelsick Road
Ambleside
Cumbria
Tel: 015394 34045
Contact: Gillean and Roger Bell
Established: 1988
Open: 9.30am to 6pm daily
Stock: Paintings, limited edition prints, wood, ceramics, jewellery, textiles, glass and sculpture by contemporary British artists and crafts people
Price: From £5 to £500. All major cards and Switch
Note: Picture framing service available

HIDE-HORN
Dixons Court
101 Lake Road
Ambleside
Cumbria LA22 0DB
Tel: 015394 33052
Fax: 015394 33052
Contact: Peter Hodgson
Established: 1985
Open: 9am to 5pm. Closed Thursday and Sunday.
Stock: Handmade horn, leather goods, saddlery, dog collars & leads.
Prices: £1 to £1,000
Note: Handmade goods and repairs.

SPIRAL POTTERY
Zeffirellis Arcade
Compston Road
Ambleside
Cumbria
LA22 9AD
Tel: 015394 41305
Fax: 015394 41773
Contact: Mike Labrum
Established: 1983
Open: 10am to 5pm
Craftworkers on site: 1
Crafts: A wide range of functional and decorative stoneware pottery.

Prices: £1 to £80. Most cards taken
Note: Visitors welcome to watch pots being made. The arcade has other shops selling high quality and unusual gifts; also an excellent cafe. Commissions are welcomed and orders can be sent by post.

HEIRLOOMS ROCKING HORSES
Dufton Hall
Dufton
Appleby-in-Westmorland
Cumbria
CA16 6DD
Tel: 01768 353353
Fax: 01768 353353
Contact: Steve Hill
Open: 8.30am to 5pm Monday to Friday. Weekends by appointment.
Craftworkers on site: 3
Craft: Rocking horses made and restored.
Prices: £300 to £2,500. All major credit cards accepted.
Note: Full visitor facilities (teashop, shop, museum and working workshop) will be open for 1997

CHATSWORTH CRAFTS
Unit 35 - Trinity Enterprise Centre
Iron Works Road
Barrow-in-Furness
Cumbria
Tel: 01229 430117
Contact: Mrs Brenda Pratt
Established: 1991
Open: 9.30am to 3.30pm. Closed for lunch 12pm to 1pm.
Crafts: Dried flower arranging. Various pottery items, garden stoneware, garden furniture, seasonal gifts. Father Xmas, Xmas trees (pottery).
Prices: £1.50 to £60. Credit cards not accepted.
Note: Craft demonstrations. This is a workshop for people with learning difficulties which produces high quality goods.

SERENDIPITY
12 Market Place
Brampton
Cumbria
CA8 1RW
Tel: 016977 41900
Fax: 01228 75657
Contact: Val Shenton
Established: 1993
Open: 9am to 5pm Monday to Saturday.

Also 10am to 4pm Sunday from May to September. Telephone call advisable
Stock: Ceramics, woodturning, garden sculptures, porcelain, hand made travel bags, handmade greeting cards, candles, pictures, découpage, glassware, high quality handcrafted giftware, miniatures, watercolours, craft kits and accessories, speciality picture frames, jewellery.
Prices: Under £5 to over £200. All major credit cards accepted.
Note: Seasonal theme exhibitions. Craft demonstrations - woodturning, canework, copperwork, picture framing, painting. All products made in Cumbria or very close. Next to tourist information centre and tea room. Ample parking

CUMBRIA CRAFT CENTRE
Linton Visitor Centre
Shaddongate
Carlisle
Cumbria
CA2 5TZ
Tel: 01228 818887
Contact: Mrs G Tyers
Established: 1994
Open: 9.30am to 5pm Monday to Saturday.
Stock: Ceramics, china, paintings, pottery, jewellery, furniture, dried flowers, soft toys, bags, jams, candles, découpage, quilts, wood turning, hand decorated mirrors, painted bargeware, slate, etc.
Prices: 99p to £400. Credit cards not accepted.
Note: Linton tweeds Heritage centre. See working hand looms. Coffee shop, picnic area. Disabled access to most areas.

EDEN VALLEY WOOLLEN MILL
Front Street
Armathwaite
Carlisle
Cumbria
CA4 9PB
Tel: 016974 72457
Fax: 016974 72457
Contact: Steve Wilson
Established: 1982
Open: 9.30am to 5.30pm daily. Sunday 12pm to 5pm
Crafts/stock: Woven textiles including floor rugs, throws, wraps, jackets, skirts, scarves, ties in wool & mohair

Prices: £2.50 to £200. Visa, Mastercard, Access accepted
Note: Hand-weaving workshops held at certain times during the year. "Made in Cumbria" member

JIM MALONE POTTERY
Hagget House - Towngate
Ainstable
Carlisle
Cumbria CA4 9RE
Tel: 01768 896444
Contact: Jim Malone
Established: 1976
Open: Showroom 9am to 6pm daily. Workshop by appointment
Craftworkers on site: 1
Crafts: High fired stoneware - functional, individual and of exceptional quality.
Prices: £5 to £250. Credit cards not accepted.
Note: All glazes are made on the premises from local minerals. The work is fired in a two chamber climbing kiln.

NORTHERN FELLS GALLERY
Orthwaite Road
Uldale
Carlisle
Cumbria CA5 1HA
Tel: 016973 71778
Contact: Mr and Mrs Wiseman
Established: 1993
Open: 10.30am to 5pm. Daily March to October. Friday, Saturday, Sunday November to March 1st. Closed January
Stock: Watercolours, oil paintings, woodturning, copperware, jewellery, ceramics, collector dolls, knitwear, cards, gifts, books
Speciality: Individually crafted items by Cumbrian artists
Prices: £1.50 to £300. Credit cards not accepted
Note: Craft demonstrations, Victorian tearoom, poetry wall. Easy access for disabled, parking, superb scenery situated in Northern Fells of Lake District National Park

THREE HORSE SHOES WOOD CRAFT
Three Horse Shoes
Irthington
Carlisle
Cumbria
CA6 4PT
Tel: 01228 75657

Fax: 01228 75657
Contact: Derek Hughes
Established: 1990
Open: 10am to 4pm. Phone call advisable. Open all year
Crafts: Decorative and functional wood turning workshop - including more than 50 different colourful timbers, burrs, nuts, roots, driftwood - from light pulls to small furnishings. Very good quality workmanship. Also some furniture restoration
Speciality: Wooden hand worked fruits
Prices: Under £5 to over £500. Visa/Access/Eurocard/MasterCard taken
Note: Demonstrations, tuition, residential activity breaks. Mobile demonstrations visiting special events, visitor centres. Crafts Council listed

THE RURAL WORKSHOPS
Lake Road
Coniston
Cumbria
LA21 8EW
Tel: 015394 41129
Contact: Nick Dune
Established: 1984
Open: 9.30am to 5.30pm
Craftworkers on site: 3
Crafts: Hand printed table linen, calendars, roller blinds, fire screens, sweatshirts, scarves, prints, cushions, all printed to order while you wait
Prices: £5 to £300. Credit cards not accepted
Note: Adjacent potter and woollen mill Small picture gallery. Coaches welcome. Disabled access to most areas.

DALTON POTTERY
8 Nelson Street
Dalton-in-Furness
Cumbria
LA15 8AF
Tel: 01229 465313
Contact: Sue/Will Thwaites
Established: 1983
Open: 10am to 5pm
Craftworkers on site: 4
Crafts: Terracotta gardenware, novelty planters, flower arrangers' products.
Prices: 20p to £30. Access and Visa accepted
Note: Please telephone before visiting to check owners are not away at a flower show

FOLD END GALLERY
Boot
Eskdale
Cumbria CA19 1TG
Tel: 01946 723213
Contact: Mr and Mrs James
Established: 1973
Open: 10am to 5.30pm all year. Closed on Mondays
Stock: Contemporary paintings (watercolours, oils, pastels), etchings, ceramics, glass, jewellery, sculpture, cards
Speciality: Lake District paintings
Prices: £2.50 to £600

HEATON COOPER STUDIO
Grasmere
Cumbria LA22 9SX
Tel: 015394 35280
Open: 9am to 6pm daily and Sunday, 12 noon to 6pm in summer. Closing at 5pm in winter.
Stock: Ceramics, sculpture, cards, books and a permanent but changing exhibition of water colours as well as artists' materials
Prices: From under £1 to over £5000. Credit cards accepted

PETER HALL & SON CABINETMAKERS
Danes Road
Staveley
Kendal
Cumbria LA8 9PL
Tel: 01539 821633
Established: 1972
Craftworkers on site: 12
Open: Monday to Friday 9am to 5pm. Saturday and Bank Holidays, showroom only 10am to 4pm
Crafts: Cabinet making, antique restoration, turned wood items and traditional upholstery
Prices: From under £1 to over £5000. Major credit cards accepted
Note: There is wheelchair access to the showroom area but there are no disabled toilets

SAGEM CRAFTS
Mint Cottage
Gilthwaiterigg Lane
Kendal
Cumbria
LA9 6NT
Tel: 01539 724707
Contact: Pauline Brocklehurst

Established: 1990
Open: Available most days but please telephone for details
Crafts: Traditional wooden board games. Hand made "Bagatelle" chess boards with drawer, and "School" a traditional Dutch family game. Fully lined wooden cases for snooker cues etc.
Prices: £70 to £125. Credit cards not accepted
Note: Commissions undertaken. Member of "Made in Cumbria"

TURNING POINT
117 Highgate
Kendal
Cumbria
LA9 4EN
Tel: 01539 732734
Contact: Hans Ullrich
Established: 1985
Open: 9.30am to 5.30pm, Monday to Saturday.
Crafts/Stock: General. Own slipware pottery.
Prices: 50p to £500

WEDGE HALL GALLERY
Windermere Road
Staveley
Kendal
Cumbria LA8 9LY
Tel: 01539 821130
Contact: Maree Braithwaite
Established: 1988
Open: 10am to 5pm Monday to Saturday. Closed Tuesdays. 12pm to 5pm Sundays. All year round.
Stock: Pottery, glassware, jewellery, turned wood, wooden toys, soft toys and dolls.
Speciality: Paintings, prints
Prices: 50p to £120. Credit cards not accepted
Note: Tea shop, tea garden - summer. Homemade cakes and scones, light lunches, etc.

GROVE HOUSE GALLERY
89 Main Street
Keswick
Cumbria CA12 5DT
Tel: 017687 73942
Fax: 017687 73942
Contact: Sarah Holmes
Established: 1979
Open: 9am to 6pm in summer, 9am to

5.30pm April to June, 10am to 5pm other months. Closed Wednesdays in January.
Stock: Hand crafted ceramics, jewellery, minerals, dolls house furniture, rugs and cushions, prints and photographs, candlesticks and candles, soft toys and pocket money items.
Speciality: Original paintings of lakeland landscapes.
Prices: £1 to £1,000. Visa, Switch, Delta JMBC accepted.
Note: Two original wells in shop. One is used as a wishing well to raise money for local charities. Grade II listed building - Regency style

LAKELAND RURAL INDUSTRIES
Grange View
Grange-in-Borrowdale
Keswick
Cumbria CA12 5XA
Tel: 017687 77226
Contact: John Barrow
Established: 1947
Open: Summer 10.30am to 4pm. Winter 10.30am to 4pm. Closed Jan.
Stock: A full range of gift and small craft work. 'Made in Cumbria' items stocked.
Speciality: Hand beaten stainless steel ware
Prices: 20p to £250. All major credit cards except Diners Club accepted
Note: Tea shop - some tables outside in summer. Cumbria Tourist Board member

LARA ALDRIDGE–GLASS DESIGN
Rose Cottage, Hallthwaites
near **Millom**
Cumbria LA18 5HP
Tel: 01229 774229
Contact: Lara Aldridge and Jon Oakley
Established: 1995
Open: 10am to 4pm Tuesday to Sunday, but please phone before travelling a long distance
Craftworkers on site: 2
Crafts: Kiln fired glass with metal inclusions; panels, bowls, cards
Prices: £4 to £300
Note: Demonstrations held, tearoom

LOUIZE SPEAKMAN DESIGNER JEWELLERY
Unit 1, Station Craft Centre
Station Road
Millom
Cumbria LA18 5AA

Tel: 01229 770000
Contact: Louize Speakman
Established: 1995
Open: 10am to 4pm Monday to Saturday. Visitors welcome by appointment
Craftworkers on site: 1
Crafts: Hand crafted silver jewellery using semi-precious stones
Prices: £20 to £500. Credit cards not accepted
Note: Commissions undertaken. Other craft units on site, main line railway platform, demonstrations, teashop, parking, gallery

ALAN STONES PRINTMAKER
The Studio
Blencarn
Penrith
Cumbria CA10 1TX
Tel: 01768 88688
Contact: Alan Stones
Established: 1984
Open: Wednesdays 1pm to 5pm
Craftworkers on site: 1
Crafts: Lithographic prints. Charcoal drawings and oil paintings
Prices: £29 to £1000. Access, Visa, Mastercard, Eurocard accepted
Note: An idyllic fellside village in the Eden Valley, situated at the foot of Cross Fell. Wheelchair access but with some difficulty

EDEN CRAFT GALLERY
St Andrew's Churchyard
Penrith
Cumbria CA11 7YE
Tel: 01768 867955
Fax: 01768 863518
Contact: Mrs Viv Marston
Established: 1982
Open: 9am to 5pm Monday to Saturday (closed Wednesdays between Christmas and Easter).
Stock: A wide variety of crafts all made locally. More than 40 different crafts are sold. Complete range of Lilliput Lane cottages.
Prices: 15p to £450. Mastercard, Visa, Amex and all major credit cards accepted.
Note: Picture framing (tapestry and all needlework) on the premises. Excellent coffee shop serving real homemade food. Mail order catalogue available. This shop is run by a local craft association.

GARDEN OF EDEN
39 Great Dockray
Penrith
Cumbria CA11 7BN
Tel: 01768 863466
Fax: 01768 863466
Contact: Jane Carroll
Established: 1995
Open: 9am to 5pm Monday to Saturday.
Half day closing Wednesday (9am to 12 noon).
Crafts/Stock: Dried and silk flowers sold separately and in arrangements, baskets and materials to do your own arrangement etc.
Speciality: Making up arrangements and weddings to order.
Prices: £3 to £65. Credit cards not accepted.
Note: Demonstrations - visitors may watch me working and ask questions - but they aren't arranged. My shop is probably the only one which specialises in such a variety of dried flowers. It is a very homely shop with bags of character.

JON CHENEY POTTERY
Low House
Thackwaite
Penrith
Cumbria CA11 0ND
Tel: 017684 86491
Established: 1989
Contact: Jon or Heather Cheney
Open: May to October, 10am to 4pm but often later so a phone call might be worth while
Craftworkers on site: 2
Craft: Well made hand thrown domestic stoneware, terracotta, kitchen and tableware.
Prices: From under £5 to over £60. Credit cards not accepted

LABURNUM CERAMICS (GALLERY)
Yanwath
near **Penrith**
Cumbria CA10 2LF
Tel: 01768 864842
Contact: Viv/Arne Rumbold
Established: 1994
Open: 10am to 4.30pm Wednesday to Sunday including Tuesday also in July/ August and extra hours/days 1st to 24th December. Ring before travel January to February
Stock: Contemporary studio ceramics –

individual and of highest quality including domestic, decorative and sculptural earthenware, stoneware, porcelain from UK and Europe. Wedding lists and gift vouchers
Speciality: Cumbrian ceramics, glass.
Prices: £2 to £800. Access/Visa/ Amex accepted.
Note: Gallery located off B5320 road between Pooley Bridge and Eamont Bridge. Sculpture garden with seats; tea or coffee. Guided tours by arrangement. Terracotta and large ware. Currently showing work of 60 ceramics and glass artists. Greatest concentration of Cumbrian potters anywhere. 6 exhibitions each year. Disabled access: please ring in advance for extra help

WETHERIGGS COUNTRY POTTERY
Clifton Dykes
near **Penrith**
Cumbria CA10 2DH
Tel: 01768 892733
Fax: 01768 892722
Contact: Lindsey Hall
Established: 1855
Craftworkers on site: 6
Open: 10am to 5.30 pm seven days a week
Craft/Stock: Terracotta and tableware made on site as well as a wide range of craft gifts – candles, cards, jewellery, figurines – in an extensive new shop and country store
Note: This old established pottery has recently been restored and remodelled with various visitor attractions including nature walks and restaurant

THE SHOP ON THE GREEN
Dent
Sedbergh
Cumbria CA10 5Q
Tel: 015396 25323
Fax: 015396 25323
Contact: Sophie Schellenberg
Established: 1990
Open: From 10am daily. In winter months during the week please phone if making a special journey.
Stock: Designer knitwear, textiles, local silversmithing, blacksmithing and pottery. Wooden crafts, games, children's toys and a large selection of cards.
Note: Designer knitwear workshop in a long tradition of knitting in Densdale.

FOREST SPINNERS
The Syke - Rusland
Haverthwaite
Ulverston
Cumbria
LA12 8JT
Tel: 01229 860255
Contact: Chris Parker
Established: 1984
Open: April to September (Sun-Fri) 11am to 5.30pm. October to March (Sun-Fri) 11am to 4.30pm.
Craftworkers on site: 2
Crafts/Stock: Hand spun articles, crocheted from wool, doghair, musk ox, mohair, camel. Handmade paper, dried flower cards, pictures, spindles, jewellery.
Prices: £1.25 to £80. Credit cards not accepted.
Note: Spinning demonstrations. A chance for visitors to try spinning, starter sessions, school visits, groups welcome. We will spin the hair from your own pet, if wanted! Free admission.

THE DOLLS HOUSE MAN
at Furness Galleries
1 Theatre Street
Ulverston
Cumbria
LA12 7AQ
Tel: 01229 587657
Contact: Anthony Irving
Established: 1984
Open: Tuesday to Friday 10am to 4pm. Saturday 9.30am to 5pm. By appointment when closed
Craftworkers on site: 2
Crafts/Stock: Dolls houses and large toys. Miniatures, wooden toys
Prices: £1 to £600
Note: Demonstrations on making dolls houses. Demonstrations and workshops on rubber stamping within the shop

IAN LAVAL FURNITURE
Meadowbank Farm
Curthwaite
Wigton
Cumbria
CA7 8BG
Tel: 01228 710409
Contact: Ian Laval
Established: 1971
Open: Most days 9am to 5.30pm or by appointment
Craftworkers on site: 1

Crafts: Traditional furniture in native oak, walnut etc.
Prices: £250 to over £6,000. Credit cards accepted
Note: Located in restored 17th century thatched cruck house with showroom. Visitors can view work in progress. Disabled access difficult

JEWELLERY BY MICHAEL KING
Oakleigh - Todd Close
Curthwaite
Wigton
Cumbria
CA7 8BE
Tel: 01228 710756
Contact: Michael/Angela King
Established: 1984
Open: 9.30am to 5pm Tuesday to Saturday (not Bank Holidays).
Crafts: Gold and silver jewellery. Selection of gem stones.
Prices: £10 to £1,000. Credit cards not accepted.
Note: This is a jewellery manufacturer - continuous demonstrations of the processes required to make jewellery by hand. Michael King is a member of the Institute of Professional Goldsmiths. Disabled access. No toilets

JOHN KERSHAW'S STUDIO POTTERY
40 Main Road
Windermere
Cumbria
LA23 1DY
Tel: 015394 44844
Contact: John Kershaw
Established: 1972
Open: 9.30am to 5.30pm, Monday to Saturday
Craftworkers on site: 1
Craft: A wide range of studio pottery is produced in a distinctive style using strong shapes and textures heavily influenced by ancient and ethnic pottery
Prices: From £12 up to £120. Major credit cards are accepted
Note: There is wheelchair access to the showroom

ALFORD POTTERY
Commercial Road
Alford
Lincolnshire LN13 9EY
Tel: 01507 463342
Contact: M. Ducos
Established: 1972
Open: 9am to 5pm Monday to Friday
Craftworkers on site: 2
Crafts: Hand thrown stoneware, pottery
giftware and tableware, terracotta garden
pots
Prices: £1 to £100. Credit cards not
accepted
Note: Working pottery with small direct
sales shop and viewing. The owners tell
us they are not seeking to be a tourist
attraction.

SANDRA BARKER BASKETMAKER
Swan Cottage
6 Gardiners Lane
Ashwell
Hertfordshire
SG7 5NZ
Tel: 01462 743009
Contact: Sandra Barker
Established: 1980
Open: Many days but please phone
before visiting as sometimes away
teaching and demonstrating
Craftworkers on site: 1
Crafts: Traditional English willow baskets
and chair seating of all types
Note: A beautiful old English village,
source of River Cam, museum, historical
church and buildings. Lovely inns etc.
Level access to the workshop no problem
for disabled

CALTHORPE WORKSHOP
Corner House, Aldborough Road
Calthorpe
Aylsham
near Cromer
Norfolk NR11 7QP
Tel: 01263 768346
Fax: 01263 768346
Contact: Paul Jackson
Established: 1986
Open: 10.30am to 5.30pm Easter to end
of September every day except Sunday.
Other times by appointment
Craftworkers on site: 1
Crafts: Porcelain sculptures, fairy tale

based storytelling/painting/designing
Prices: 50p to £10. No credit cards
Note: Workshop open to view. Two
exhibitions a year – Xmas and Summer

THE OLD SURGERY
Doctors Corner, Aldborough
Aylsham
near Cromer
Norfolk NR11 7NR
Tel: 01263 768007
Contact: Ken and Karen Fidgen
Established: 1975
Open: Always open at reasonable times
but phone call first would be appreciated
Craftworkers on site: 2
Crafts: Silk screened greetings cards,
hand painted furniture
Prices: £1.50 to £100. No credit cards
Note: Screen printing by hand usually in
progress. Located on edge of very
picturesque village – excellent pub meals

WAVENEY ENTERPRISES
13 Smallgate
Beccles
Suffolk NR3 3AB
Tel: 01502 716065
Contact: Mr. B.K. Parker
Established: 1983
Open: 9am to 4pm, Monday to Friday
Stock: Leatherwork, woodwork, printing,
weaving, toys, wooden signs
Speciality: Signs & personalised products
Prices: 50p to £150. No credit cards
Note: Waveney Enterprises is a registered
daycare charity for up to 18 people with
special needs making and selling goods

**BARLEYLANDS GLASSWORKS AND
CRAFT CENTRE**
Barleylands Road
Billericay
Essex CM11 2UD
Tel: 01268 293000
Contact: Bill Rhodes
Open: 10am to 4pm daily . Not Monday
Craftworkers on site: 2
Crafts: Glassware designed and made by
Barleylands team
Prices: From under £5 to over £70. Credit
cards accepted
Note: Visitors can view glassblowing.
Attractions on the site include restaurant,
farm museum and other craft workshops

ROSALIND BATTSON
Meadow View
High Street
Souldrop
near **Bedford**
Bedfordshire
Tel: 01234 781472
Contact: Rosalind Battson
Craftworkers on site: 1
Open: Most weekdays normal business
hours but advisable to phone first
Crafts: Stained glass objets d'art. Mirrors,
lampshades (panel or repro Tiffany),
jewellery boxes, clocks, planters and
terraria and light catchers
Note: Commissions undertaken. Nearby
attractions include RSPB Nature Reserve,
Sandy, Woburn Safari Park

BLYTHBURGH POTTERY
Chapel Lane
Blythburgh
Suffolk
IP1 9LW
Tel: 01502 478234
Contact: Dorothy Midson
Established: 1989
Open: 11am to 5.30pm
Craftworkers on site: 1
Crafts: Handmade one-off pots
Prices: £4 to £200. No credit cards
Note: Blythburgh Church opposite and
river walks. Suffolk's smallest working
pottery (12ft square)

HANSEL UND GRETEL
1 The Green
Finchingfield
Braintree
Essex
CM7 4JX
Tel: 01371 811211
Fax: 01440 788276
Contact: Christa Ginn
Established: 1992
Open: 6 days a week 11am to 4.30/5pm.
Closed Mondays
Stock: Doughcraft and many cuckoo
clocks. Hand painted items along with
hand painted high quality furniture and
beds. Also Tyrolean style interior design
service. All items from South Germany,
Austria or Switzerland
Prices: 35p to £1,500. Credit Cards not
accepted
Note: Finchingfield is a picturesque
village with pubs and teashops

BROMHAM MILL AND ART GALLERY
Bridge End
Bromham
Bedfordshire
MK43 8LP
Tel: 01234 824330
Fax: 01234 228921
Established: 1986
Open: March to October Wednesday to
Saturday 12 noon to 4pm. Sunday and
Bank Holidays 10.30am to 5pm
Stock: Majority of work by gallery
exhibitors. Changing monthly exhibitions,
regular fabric focus and emphasis on
crafts. Cabinet pieces supplied by Crafts
Council recognised gallery
Prices: £2.50 – £10,000. No credit cards
Note: Quality tea, coffee and snacks
within working watermill with flour
milling demonstrations. Please phone for
a current programme of exhibitions and
events as the site is actively being
developed

CAMBRIDGE CONTEMPORARY ART
6 Trinity Street
Cambridge
CB2 1SU
Tel: 01223 324222
Contact: Denise Collins
Established: 1990
Open: Monday to Saturday 9am to
5.30pm
Stock: Original limited edition prints,
paintings, sculpture and crafts. High
quality ceramics and glass, metalwork,
furniture, mirrors. Bespoke framing
service
Prices: A wide range up to £2000. Credit
cards accepted
Note: Monthly exhibitions

MOMO THE COBBLER
8 Cobbe Yard
Napier Street
Cambridge
Cambridgeshire
CB1 1HR
Tel: 01223 358209
Contact: Momo (Mr. A.M. Iraninejad)
Established: 1981
Open: 9am -5 pm. Monday to Saturday
Craftworkers on site: 1
Crafts: Leatherware. Shoes and bags
made to measure. Leatherware repaired
Prices: £1 to £300. Credit cards not
accepted

RICHARD WILSON
The Workshop
70A Hartington Grove
Cambridge
Cambridgeshire CB1 4UB
Tel: 01223 566638
Fax: 01223 566639
Contact: R. Wilson
Established: 1989
Open: 9am to 1pm - 2pm to 5pm
Craftworkers on site: 1
Crafts: Bows for all string instruments
Prices: £450 to £1800

THE COGGESHALL POTTERY
49 West Street
Coggeshall
Essex CO6 1NS
Tel: 01376 561217
Contact: Peter Turner
Established: 1983
Open: Wednesday to Sunday 11am to
5pm
Craftworkers on site: 2
Crafts: Handmade pottery specialising in
indoor water gardens and fountains.
Some domesticware in stoneware.
Terracotta garden pots
Prices: £1.50 to £130
Note: Access to only the shop is available
to the public. Workshop by arrangement

PAM SCHOMBERG GALLERY
12 St John's Street
Colchester
Essex CO2 7AN
Tel: 01206 769 458
Contact: Pam Schomberg
Established: 1992
Open: Monday - Saturday 10.30am to
5pm
Stock: Very high quality domestic pottery,
studio ceramics, wood, metal, textiles,
glass, jewellery, toys
Prices: From under £5 to £1000
Note: Independent gallery promoting
British contemporary applied art and
craftmanship

DEDHAM ART & CRAFT CENTRE
High Street
Dedham
Essex CO7 6AD
Tel: 01206 322666
Fax: 01206 322644
Contact: Sandy Collier & Jeremy Parkin
Craftworkers on site: about 20

Open: Daily 10am to 5pm. Closed
Christmas and Mondays from January to
March. Please telephone for opening
times of toy museum
Crafts/stock: Collective of artists,
candlemaker, tie maker and ceramic
figurine artist. Dried flowers, hats, clothes,
painted glassware, pottery, hand
decorated silk scarves, murals, waistcoats,
wood turning, cushions, toys, jewellery,
books, cards and papers
Prices: £1 to £500. Credit cards taken
Note: Wholefood family restaurant, tea
room, small toy museum, set in beautiful
Dedham Village famous for its connec-
tions with John Constable.

THE PARTICULAR POTTERY & GALLERY
West Church Street, Kenninghall
near **Diss**
Norfolk NR16 2EN
Tel and Fax: 01953 888476
Contact: Michelle Walters
Established: 1988
Craftworkers on site: 1
Open: Tuesday to Saturday 9am to 5pm.
Also open Bank Holiday Mondays
Crafts/Stock: Hand thrown porcelain/
ceramics, framed prints and special
exhibitions of original hand made crafts at
Easter, autumn and at Christmas.
Prices: £10 to £200. Most major credit
cards accepted
Note: The pottery is situated in a
sympathetically restored Baptist Chapel
(1807). Conservation village. Free colour
brochure available. Disabled access is
limited – downstairs workshop accessible
but space limited, please telephone first

TRAIDCRAFT ETC
4 Church Street
Diss
Norfolk IP22 3DD
Tel: 01379 650067
Contact: Irene Sayer
Established: 1990
Open: 9.15am to 5pm. Closed 2pm
Tuesdays apart from holidays
Stock: Third World crafts, jewellery, re-
cycled paper and cards. T-shirts,
Romanian embroidery.
Speciality: Cards, earrings and incense
Prices: Few pence to £25. No credit cards
Note: Wheel and pushchair friendly. Tea
& coffee shop. Fairly traded goods at fair
prices. Environmentally friendly goods

FINISHING TOUCHES
Sparrow Hall Farm
Leighton Road, Edlesborough
near **Dunstable**
Bedfordshire LU6 2ES
Tel: 01525 222636
Fax: 01525 222636
Contact: Celia Kilfoyle/Rosemary Pratt
Established: 1987
Open: 10am to 5pm Tuesday to Sunday
incl. Closed 1 pm to 2pm Tuesday to
Friday. Not Mondays except Bank Holidays
Stock: Wide selection of gifts, crafts and
prints, dried and silk flowers
Prices: £1.99 to £100. No credit cards
Note: Small tea shop - home made
cakes. Wide selection of pine and cane
furniture in other barns

NEW FROM OLD
The Engine House, Little Ouse
Ely
Cambridgeshire CB7 4TG
Tel: 01353 676227
Fax: 01353 676313
Contact: Rick Forward
Established: 1991
Open: Most days but please phone first
Craftworkers on site: 1-2
Crafts: Traditional country furniture and
kitchens handcrafted from reclaimed and
sustainable timber
Prices: £15 upwards. No credit cards
Note: Stock items include: wall shelves,
corner shelves, painted cupboards,
magazine racks etc. Furniture made to
commission. No disabled special facilities

TWENTYPENCE POTTERY
Twentypence Road
Wilburton
Ely
Cambridgeshire CB6 3RN
Tel: 01353 741353
Contact: Chris Arnold and Pete Dodge
Established: 1981
Open: 9am to 6pm daily including
Saturday and Sunday (occasionally closed
Wednesday)
Craftworkers on site: 2
Crafts: Stoneware and terracotta hand
thrown pottery. Turned wood, furniture,
violins
Prices: £1 to £75. No credit cards
Note: Work in progress always visible.
Individual orders undertaken in pottery or
furniture

CHURCH COTTAGE POTTERY
Wilby
near **Eye**
Suffolk IP21 5LE
Tel: 01379 384253
Contact: Tom and Heather Baker
Established: 1987
Open: Open all year (except January)
10am to 6pm. Sunday 2pm to 6pm.
Closed Wednesdays
Craftworkers on site: 2
Crafts: Handmade domestic, garden and
decorative pots in terracotta & stoneware.
Bargeware & pyrographed wood
Prices: 95p to £125. Credit cards not
accepted. Cheque with card accepted
Note: Craft demonstrations by arrange-
ment. Teashop and garden open for
homemade savouries and cakes. Pottery
in 18th century thatched cottage on
B1118 road - Wilby is a mile south of
Stradbroke - opposite church. Car parking
available at rear. Access for disabled

RYBURGH POTTERY
1 May Green
Little Ryburgh
Fakenham
Norfolk NR1 0LP
Tel: 01328 829543
Contact: Stephen Parry
Established: 1984
Open: Most days but advisable to
telephone if making a special journey
Craftworkers on site: 2
Crafts: Pots (wood fired stoneware and
porcelain)
Prices: £3 to £150

HEMSWELL CRAFT CENTRE
Lindsey House
Hemswell Cliff
Gainsborough
Lincolnshire DN21 5TH
Tel: 01427 667066
Fax: 01427 667178
Contact: Dave Green
Established: 1994
Open: 10am to 5pm daily
Stock: Products from up to 60 craft
producers and artists plus other design-
led products
Prices: £1 to £200. Mastercard/Visa/
Switch accepted
Note: Coffee shop, craft workshops,
adjacent to largest antique centre in U.K.
Disabled facilities

MANOR STABLES CRAFT WORKSHOPS
Fulbeck
Grantham
Lincolnshire
NG32 7JN
Tel: 01400 272779
Contact: Anne Wood
Established: 1987
Open: 10.30am to 4.30pm Tuesday to Sunday and Bank Holiday Mondays. Closed Christmas
Crafts/stock: Weaving, needlework, dried flowers, jewellery, pottery, glass, toys, books, saddlery, decorative ironwork, wood turning, furniture, paintings, cards
Speciality: Textiles, dried flowers, saddlery, ironwork
Prices: 25p to £500. Access/Visa/Mastercard/Eurocard
Note: Working craft units where commissions are undertaken. Classes taught and talks by arrangement. Daily demonstrations of speciality crafts. Teashop and ice cream parlour

OASBY POT SHOP
Oasby
Grantham
Lincolnshire
NG32 3NB
Tel: 01529 455234
Contact: John I. Evans
Established: 1988
Open: 10am to 5pm. Closed Wednesdays
Stock: Pottery from some 40 potters gives an exceptionally wide choice of design and decoration, from studio pottery to pots for everyday use
Speciality: British, hand thrown and decorated craft pottery
Prices: £2 to £100. Credit cards not accepted
Note: Garden, small range of complementary gifts

CASTLE HEDINGHAM POTTERY
St James Street
Castle Hedingham
near **Halstead**
Essex
CO9 3EW
Tel: 01787 460036
Contact: Mr. & Mrs. John West
Established: 1978
Open: 9am to 5.30pm 6 days a week. Closed Mondays

Craftworkers on site: 2
Crafts: Domestic pottery, garden pots, commemorative pots, house signs
Prices: 50p to £50. No credit cards
Note: Talks and demonstrations. One-day pottery courses (no experience required). Coffee, lunch and tea provided. Bed and breakfast available. Potters wheels made to order

MILLHOUSE POTTERY
1 Station Road
(Opp. Duke William P.H.)
Harleston
Norfolk
IP20 9ES
Tel: 01379 852556
Contact: Alan & Ann Frewin
Established: 1965
Open: 10am to 5.30pm daily
Craftworkers on site: 2
Crafts: High fired earthenware, decorated baking dishes, large and small jugs, mugs, bowls, etc. Large range of slipware, tin glaze. Decorative garden pots
Prices: 60p to £450. Credit cards not accepted
Note: Showroom, garden and courtyard. Free tea and coffee. Shrubs and plants

MADE IN CLEY
High Street
Cley-next-the-Sea
near **Holt**
Norfolk
NR25 7RF
Tel: 01263 740134
Contact: Barbara Widdup
Established: 1984
Craftworkers on site: 6
Open: October to June daily 10am to 5pm except Sunday 11am to 5pm. Closed all day Wednesday. July, August, September daily 10am to 5pm ,Sunday 11am to 5pm
Crafts: Domestic stoneware, pottery, contemporary jewellery, sculpture in marble and other stones. All made on premises
Prices: £5 to £1,000. Visa/Access/Amex/Diners Club
Note: Cley-next-the-Sea is a lovely coastal village with windmill, various shops and tearoom. Members of Craft Potters Association, Norfolk Contemporary Craft Society and East Anglian Potters Association

WROXHAMS BARNS LTD
Tunstead Road
Hoveton
near Norwich
Norfolk NR12 8QU
Tel: 01603 783762
Fax: 01603 783911
Contact: Diane Harber
Established: 1983
Open: 10am to 5pm all year. Closed
December 25th and 26th
Craftworkers on site: 14
Crafts/Stock: Excellent range of gifts and
crafts for all ages, tastes and "pockets"
Prices: £1 to £100
Note: Tea room, coffee shop, junior farm,
Williamson traditional fair. Excellent day
out. Entry and car parking free. Thirteen
units are accessible to wheelchairs using
ramp. Tearoom, coffee shop and toilet
also accessible. Pathways are gravelled

SOMERSHAM POTTERY
3 and 4 West Newlands Ind. Estate
Somersham
Huntingdon
Cambridgeshire PE17 3EB
Tel: 01487 841823
Fax: 01487 841042
Contact: Stuart Marsden
Established: 1978
Open: 9am to 5pm Monday to Saturday.
1pm to 5pm Sunday
Craftworkers on site: 4
Crafts: Handmade, wheel thrown pottery.
Cards. Crystal glaze porcelain and full
range of domestic ware, house plaques,
commemorative plates, trophies for clubs
Prices: £1 to £500. Credit cards not taken
Note: Wheel demonstrations, visitors
allowed to have a go. Commissions
accepted

CORNCRAFT
Monks Eleigh
Ipswich
Suffolk IP7 7AY
Tel: 01449 740456
Fax: 01449 741665
Contact: Win Gage
Established: 1973
Open: All year. Monday to Saturday
10am to 5pm. Sunday 11am to 5pm
Stock: Locally made crafts, pottery,
jewellery, pot pourri, aromatic products,
cards and stationery products
Speciality: Dried flowers and corn dollies

Prices: £1.50 to £100. Visa, Mastercard,
Delta, Connect
Note: Tea shop. Craft demonstrations for
groups by appointment. Beautifully
located on a farm

CAITHNESS CRYSTAL VISITOR CENTRE
10-12 Paxman Road
Hardwick Industrial Estate
King's Lynn
Norfolk PE30 4NE
Tel: 01553 765111 **Fax:** 01553 767628
Contact: D. Johnson
Established: 1987
Open: 6 days January to Easter. 7 days
Easter to Christmas.
Craftworkers on site: 3
Stock: Factory shop selling giftware,
stemware, glass animals, paperweights.
Prices: £3.99 to £190. Major credit cards
Note: Coffee, light lunches and snacks

CORN DOLLY CRAFTS & GIFTS
7 Peacock Mews
Leighton Buzzard
Bedfordshire LU7 8JH
Tel: 01525 853096
Contact: Mrs. C.A. Scholes
Established: 1991
Open: 9.30am – 5pm (Thurs 10am –2pm)
Stock: High quality UK crafts, candles,
wrought iron, silver, ceramics, collectors'
teddy bears, local artists, turned wood
Speciality: Candles, teddies, art.
Prices: £1 to £250. Mastercard/Delta,/
Access/Visa/Eurocard
Note: Exhibitions, craft courses & dem-
onstrations all year. Tea shop next door

ALDRINGHAM CRAFT MARKET
Aldringham
near **Leiston**
Suffolk IP16 4PY
Tel: 01728 830397
Contact: Godfrey Huddle
Established: 1958
Open: Monday to Saturday 10am to
5.30pm. Sunday 10am to 12pm (Spring &
Summer) 2pm to 5.30pm (all year)
Stock: Ceramics, both studio and
domestic, terracotta garden pots, turned
wood, original paintings, etchings, prints,
toys, kites, ladies' clothes, gifts, cards,
books, jewellery, toiletries, etc.
Prices: £1 to £600. Major credit cards
Note: Many exhibitions each year.
Coffee shop and children's play area.

BEADAZZLED
14 The Mall
Lincoln
Lincolnshire LN2 1HX
Tel: 01522 522116
Fax: 01522 790556
Contact: Lorraine and Graham Cadel
Established: 1991
Open: 9.30am to 5pm
Craftworkers on site: 3
Crafts/stock: Quality handcrafted
jewellery, beads and findings
Prices: 1p to £70. No credit cards
Note: Free instruction given where
necessary. Talks and demonstrations
given with hands-on jewellery making.
Jewellery co-ordinated for both style and
colour for special occasions.

COBWEB CRAFTS
The Old School
Cadney Road, Howsham
Lincoln
Lincolnshire LN7 6LA
Tel: 01652 678761
Contact: Kevin P. Burks
Established: 1976
Open: 9am to 5pm Monday to Saturday.
By appointment after hours and Sundays
Craftworkers on site: 4
Crafts: English hardwoods (oak) cabinet
and furniture makers
Prices: £150 to £10,000

THE STABLES STUDIO
94 High Street, Martin
Lincoln
Lincolnshire LN4 3QT
Tel: 01526 378528
Contact: Mrs. Jo Slone
Established: 1995
Open: 10am to 6pm Wednesday to
Sunday. Closed Monday and Tuesday
Stock: Contemporary ceramics, sculpture,
paintings, wood, individual knitted hats,
African art, cards, gift, photography etc.
Prices: £2 to £300. Credit cards not
accepted
Note: Sculpture garden and patio teas
(weather permitting)

THE STUDIO AT STONECOTT
Bar Lane
Waddington
Lincoln
Lincolnshire
LN5 9SA

Tel: 01522 722297
Contact: Maggie Betley and Tracy Wright
Established: 1988
Open: 11am to 4.30pm Wednesday to
Saturday. Other times by appointment
Craftworkers on site: 2
Crafts/stock: Original handcrafted
ceramic gift/art work, unusual designs/
colours. Gargoyles, clocks, vases, oil
burners, sculpted birds, jewellery, beads,
buttons and more. Wooden sculpture,
water colours, dough designs, pyrography, bronze casts, handmade cards, bees
wax candles, barge work etc.
Prices: £2.50 to £200. Visa/Mastercard
Note: Located in old part of attractive
village. Free parking and disabled toilets
opposite

TIMBERLAND ART & DESIGN
12 Church Lane
Timberland
Lincoln
Lincolnshire
LN4 3SB
Tel: 01526 378222
Contact: Janet Crafer and Jonathan
Korejko
Established: 1990
Open: 10am to 5pm Tuesday to Friday.
Sunday 1pm to 4pm. Closed first Tuesday
of every month
Craftworkers on site: 2
Crafts: Printmaking and tapestry weaving
Prices: £5 to £5,000. Credit cards not
accepted
Note: Studio/gallery in converted
Victorian chapel. Picturesque village. Full
access to wheelchair users

MANNINGTREE POTTERY AND GALLERY
18-20 High Street
Manningtree
Essex
CO11 1AG
Tel: 01206 396380
Fax: 01206 392391
Contact: Mike Goddard
Established: 1986
Open: 9.30am to 5.30pm
Craftworkers on site: 3
Crafts: Stoneware and domestic ware
Prices: £5 to £500. Credit cards not
accepted.
Note: Open workshop and demonstra-
tions possible

THE POTTERY
Narborough
Norfolk
PE32 1TE
Tel: 01760 337208
Contact: John and Kate Turner
Established: 1965
Open: Most days but please telephone first
Craftworkers on site: 2
Crafts: Domestic stoneware, garden pots, ceramic sculpture
Note: Next door to pub with fishing. Caravan site with pool. Cheerful workshop - visitors always welcome

CAT POTTERY
1 Grammar School Road
North Walsham
Norfolk
NR28 9JH
Tel: 01692 402962
Contact: Ken Allen
Established: 1955
Open: 9am to 5pm Monday to Friday 11am to 1pm Saturday
Craftworkers on site: 5
Crafts: Slip cast pottery. Cats with glass eyes
Prices: £4 to £50. Credit cards not accepted
Note: This is a pottery workshop. Collection of railway memorabilia and other curiosities

COLTISHALL CRAFTS
Church Street
Coltishall
near **Norwich**
Norfolk
NR12 7DW
Tel: 01603 737549
Contact: Elizabeth Tooth
Established: 1972
Open: Easter to 30th Sept 7 days from 10am to 6pm. 1st Oct to 30th Dec 10am to 4pm. Closed Mondays
Stock: Quality local Norfolk crafts. Furniture, watercolours, marquetry, ceramics, pottery, needlecraft, wood turning, children's and soft toys
Prices: 99p to £2,500. Visa/Mastercard accepted
Note: Located 300 yards from Coltishall Staith on the Norfolk Broads. Bed and breakfast available in 17th century country home which adjoins the property

JANE'S CRAFT SHOP
Taverham Garden Centre
Fir Covert Road, Taverham
Norwich
Norfolk NR8 6HT
Tel: 01603 260942
Contact: Mrs. J.M. Fleming
Established: 1989
Open: Monday to Saturday 10am to 5pm. Sunday 10.30am to 4.30pm
Stock: Knitwear, jewellery. A full range of gifts, collectables and small craft work
Speciality: Watercolour paintings
Prices: £1 to £300. Visa/MasterCard/ Diners Club
Note: Set in garden centre complex, tea shop opposite. Embroidery and needle-craft stockist, dolls house and miniatures, kitchen shop all close by. Ramped premises for disabled

RAINDROP
9A, St Mary's Works
St. Mary's Plain, Duke Street
Norwich
Norfolk NR3 3AF
Tel: 01603 767653
Contact: Helen Howes
Established: 1978
Open: 10am to 6pm Tuesday to Friday. Closed Bank Holidays
Craftworkers on site: 1
Crafts/Stock: Kites, knitwear, mobiles, clocks, jugglers' balls, frisbees, flags, hats, waistcoats, badges, boxes, turning clocks, banners, buttons, cards
Prices: 10p to £500. Credit cards not accepted
Note: World famous kitemaker. Commissions welcome. 1 hour free parking. Cafe next door. Wheelchair access.

THE GOWAN GALLERY
3 Bell Street
Sawbridgeworth
Hertfordshire
CM21 9AR
Tel: 01279 600004
Contact: Joanne Gowan
Established: 1987
Open: 10am to 5pm Tuesday to Saturday (closed one week Christmas to New Year). Also open most Mondays
Stock: High quality jewellery, glass, ceramics, metalwork, wood, limited edition etchings all by British designer makers and artists

Speciality: Designer jewellery in precious metals
Prices: £10 to £2,000. Access/Mastercard/Visa
Note: Precious designer jewellery by Joanne Gowan made on the premises. Crafts Council selected since 1990

JONATHAN KEEP POTTERY
31 Leiston Road
Knodishall
Saxmundham
Suffolk IP17 1UQ
Tel: 01728 832901
Contact: Jonathan Keep
Established: 1986
Open: 9.30am to 5.30pm Monday to Saturday
Craftworkers on site: 1
Crafts: Handmade decorated domestic pottery
Prices: £3 to £60. Credit cards not accepted
Note: Member of Suffolk Craft Society and East Anglian Potters Association.

"BY GEORGE!" CRAFT ARCADE
23 George Street
St. Albans
Hertfordshire AL3 4ER
Tel: 01727 853032
Contact: D.J. Pyne
Established: 1986
Open: 9.30am to 5pm Monday to Saturday - 1pm to 5pm Sunday - throughout the year
Crafts/stock: Wood turning, glass engraving, kites, collector bears and dolls, ceramics, flowers, millinery, jewellery, pottery, dolls houses, traditional toys, stained glass, pewter, book binding, artists, collector car models, silversmith, furniture maker, needlework
Prices: £2.50 to £500. Credit cards not accepted
Note: Part of the 'By George 'complex of craft arcade, antique centre & restaurant

THE PEAROOM CENTRE FOR CONTEMPORARY CRAFTS
Station Yard
Heckington
Sleaford
Lincolnshire NG34 9JJ
Tel: 01529 460765
Contact: Caroline Jarvis
Established: 1989

Open: 10am to 5pm, Monday to Saturday and Sunday 12pm to 5pm
Crafts/stock: Domestic pottery, wood, metal, textiles, glass, studio ceramics, jewellery, knitwear, baskets
Prices: From under £5 to over £200
Major credit cards accepted
Note: Contemporary crafts centre with shop, two galleries, workshops, commissioning centre and cafe

SUTTON (WINDMILL) POTTERY
Church Road, Sutton
Stalham
Hoveton, near Norwich
Norfolk NR12 9SG
Tel: 01692 580595
Contact: Malcolm Flatman
Established: 1977
Craftworkers on site: 1
Open: Weekdays usually 9am to 1pm and 2pm to 6pm. Weekends by arrangement
Crafts: Handmade reduction stoneware pottery in range of colours, including dinner services, tableware, kitchenware, decorative items, lamps, house name plaques
Prices: 90p to £150. Credit cards not accepted
Note: Items made to customers' specifications. Visitors are welcome to watch work in progress. All items for sale are made on the premises. Rural location but close to A149 and the Norfolk Broads. Wheelchair access fairly good – one level workshop with short ramps – better with advance notice. Every assistance given

TORKINGTON GALLERY
38 St. Peter's Street
Stamford
Lincolnshire
PE9 2PF
Tel: 01780 62281
Contact: Will and Chris Illsey
Established: 1987
Open: 10am to 5pm Monday to Saturday
Craftworkers on site: 2
Crafts/stock: Ceramics, jewellery, carved and turned wood, hand painted silks, glass and some paintings and prints
Speciality: Ceramics
Prices: £5 to £500. Visa/Mastercard/Switch accepted
Note: Disabled access: difficult for wheelchairs, 2 steps to entrance

WOODPECKERS
The Thatched Cottage
High Street
Stalham
Hoveton, near Norwich
Norfolk
NR12 9BD
Tel: 01692 582979
Contact: Linda Blair/Chris Sparks
Open: Monday, Tuesday, Thursday, Friday, Saturday 9am to 5pm. Wed 9am to 1pm
Stock: Good range of glass, pottery, bargeware, wooden toys and puppets
Speciality: Rocking horses and dolls houses
Prices: 99p to £1,200. Credit cards not accepted
Note: Children's workshop 2 nights a week and boys woodwork club 1 night a week aimed at 7-12 year-olds. Within 5 minutes walking distance from the Norfolk Broads and boat yard

CARTERS TEAPOT POTTERY
Low Road
Debenham
Stowmarket
Suffolk
IP14 6QU
Tel: 01728 860475
Fax: 01728 861110
Contact: Tony Carter
Established: 1978
Craftworkers on site: 15
Crafts: Collectable teapots
Prices: £4.50 to £50. Credit cards accepted
Note: Pottery of world repute making unusual teapots. Tea and coffee available weather permitting, in summer months only.

WHISTLING MOUSECRAFTS
Norwich Road (A140)
Stonham Parva
Stowmarket
Suffolk IP14 5JT
Tel: 01449 711000
Contact: Cora Pullen
Established: 1989
Open: Wednesday to Sunday and Bank Holiday Mondays 10am to 5.30pm (closed January)
Stock: Collectors porcelain dolls, pottery, cards and stationery, painting and prints, aromatherapy oils and burners, rubber

stamping, jewellery, traditional toys
Prices: under £1 to £100. No credit cards
Note: Tearoom, homemade cakes, scones, etc. Fresh tea and filter coffee. Seating on patio in good weather. Off road parking

FAYS
Well Lane
Clare
near **Sudbury**
Suffolk CO10 8NH
Tel: 01787 277646
Fax: 01787 277182
Contact: Liz Fay
Established: 1986
Open: Monday to Friday: 8.30am to 5.30pm, Saturday: 8.30am-5pm. Sunday: 9am to 5pm March to October. 9am to 4pm November to February
Stock: From over 30 craftspeople – wood, glass (painted), silk embroidery, jewellery, cards, flowers, mohair, pictures, etc.
Prices: 50p to £100. No credit cards
Note: Situated in picturesque town with castle, country park, traditional Suffolk architecture and wool church. Specialist delicatessen, wine shop and florists

MUCH HADHAM FORGE
High Street
Much Hadham
Ware
Hertfordshire
Tel: 01279 843856
Fax: 01279 843675
Contact: Ted Horton
Established: 1990
Open: Tuesday to Saturday incl. 8.30am to 5.30pm
Craftworkers on site: 1
Crafts: Decorative wrought ironwork (speciality - curtain poles and fittings)
Note: Forge museum (80p entry). Victorian garden. Blacksmith demonstrations to parties for £15 per hour. Forge is in 16th century building - has been a forge since 1811

TRADING PLACES GALLERY
11 New Road
Ware
Hertfordshire
SG12 7BS
Tel: 01920 469620
Fax: 01920 463003
Contact: Mrs. K . Walden

Established: 1988
Open: Tues and Sat 11.30am to 4.30pm
or by arrangement. Exhibitions and
Christmas, Tue, Fri, Sat, Sun only
Stock: Blown and kiln glass, studio
ceramics, painting, some jewellery and
bronzes
Speciality: Glass/paintings
Prices: £20 to £2,000. No credit cards
Note: When main gallery is closed
artwork can be viewed and purchased at
Green Park Hotel, Dane End

LANNOCK POTTERY
Weston Barns
Hitchin Road
Weston
Hertfordshire SG4 7AX
Tel: 01462 790356
Fax: 01462 790356
Contact: Andrew Watts
Established: 1980
Open: Monday to Friday 10am to 5pm.
Sunday 11am to 5pm
Craftworkers on site: 5
Crafts/Stock: Thrown stoneware pottery,
cards, baskets, candles
Prices: £4 to £150. Visa/Mastercard etc.
Note: View potters at work. Excellent pub
opposite

BUTLEY POTTERY
Butley Barns
Mill Lane
Butley
near **Woodbridge**
Suffolk
IP12 3PA
Tel: 01394 450785
Fax: 01394 450843
Contact: Honor Hussey
Established: 1978
Open: 10.30am to 5pm daily April to
September. October to March 11am to
4pm (closed Mon and Tues). January
closed. Phone to confirm opening times
during winter
Craftworkers on site: 2
Crafts/stock: Pottery, sculpture, coppiced
ash furniture, modern paper moon lights
Prices: £5 to £800.Visa/Euro/MasterCard/
Access
Note: Gallery, tearoom, thatched and
weatherboarded barns in rural setting.
Bed and breakfast. Other craft workshops.
Changing exhibitions. Short courses in
arts and crafts

stop press

STUDIO GALLERY
The Old Dairy
Elton
Peterborough
Cambridgeshire
PE8 6SQ
Tel: 01832 280023
Fax: 01832 280080
Contact: Glenys Carter
Established: 1995
Open: Monday to Friday, 10am to 5pm
and Sunday afternoons in summer
Stock: Contemporary arts and crafts
including ceramics, metal work and glass
as well as etchings, watercolours and
prints.
Prices: £25 to £500. Major credit cards
are accepted
Note: Close to the famous Bressingham
plant centre. Disabled access is possible

DERWENT CRYSTAL
Shawcroft
Ashbourne
Derbyshire DE6 1GH
Tel: 01335 345219
Fax: 01335 300225
Contact: Carole Doggett
Established: 1980
Open: Monday to Saturday 9am to 5pm
Craftworkers on site: 12
Crafts: English full lead crystal glass
tableware and giftware. Engraving service
available. Always special offers
Prices: 50p to £100. Visa/Access/Amex/
Diners
Note: Viewing of glassmaking and
decorating. Phone for exact times

ST. JOHN'S STUDIO
50B St Johns Street
Ashbourne
Derbyshire DE6 1EL
Tel: 01335 345965
Contact: Sue Riley and Richard Gregory
Established: 1994
Open: 9am to 5pm daily. Closed
Mondays but open Sundays 11am to 4pm
Crafts/stock: Ceramics produced by
owner. Bead jewellery plus beads and
findings, pictures, prints, ceramic
restoration, turned wood plus other local
crafts
Speciality: Ceramics hand decorated and
ceramic restoration
Prices: £1 to £500. No Credit cards
Note: Working studio where customers
can see how it's done. Classes are held 3
times a week in decorating ceramics

FERRERS CRAFT CENTRE AND GALLERY
Staunton Harold
Ashby-de-la-Zouch
Leicestershire
LE65 5RW
Tel: 01332 863337
Contact: Jayne McKay/Jackie Blunt
Established: 1990
Craftworkers on site: 14
Open: All year. Tuesday to Sunday 11am
to 5pm. Closed Monday
Crafts/Stock: Wide range of high quality
British craftsmen made work on three
floors including ceramics, glass, furniture,
leather, textiles and ironwork. Gallery
shows work from best around the UK

Speciality: Furniture
Prices: £3 to £1,000. Gallery accepts
Switch/Barclaycard/Visa/Access -
individual workers may not
Note: Set in Georgian stableblock, now a
working craft centre with tearooms in
beautiful estate. Entertainment in courtyard Easter to October every Sunday.
Admission on Sundays £1 adults, children
free. Crafts Council selected gallery

CATAWARE
7 Hebden Court, Matlock Street
Bakewell
Derbyshire DE45 1EE
Tel: 01629 815171
Fax: 01629 814419
Contact: Angus Stokes
Established: 1992
Open: Approx. 10am to 5pm Monday to
Saturday. 12.30pm to 5pm Sunday
Stock: Wide selection of crafts from many
countries all on a cat theme
Prices: 30p to £50. All major credit cards
and debit cards accepted
Note: Products available by mail order
through a CATalogue and MEWSletter

HOOK NORTON POTTERY
East End Farmhouse
Hook Norton
Banbury
Oxfordshire
OX15 5LG
Tel: 01608 737414
Fax: 01608 730442
Contact: Russell Collins
Established: 1969
Open: Monday to Friday 9am to 5pm
Crafts: Pottery and ceramics
Speciality: Catering cookware
Prices: £2 to £100. Visa/Access accepted

MAMBLE CRAFT CENTRE
Church Lane
Mamble
near **Bewdley**
Worcestershire DY14 9JY
Tel: 01299 832834
Fax: 01299 832132
Contact: Catherine Duggins
Established: 1995
Open: Tuesday to Saturday 10.30am to
5pm. Sunday 11.30am to 5.30pm. Closed
Mondays

Craft/stock: Wood, pottery, glass, jewellery, textiles, paintings, cards, toys
Prices: 10p to £325. Visa/Access
Note: Tearoom, picture framers and gallery, watercolour artist, gift shop, craft gallery, craft demonstrations and courses held. Stunning views from the Teme Valley to the Black Mountains in Wales. Great local pubs

GOHIL EMPORIUM
381 Stratford Road
Sparkhill
Birmingham
West Midlands B11 4JZ
Tel: 0121 771 3048
Fax: 0121 772 3844
Contact: Mr. and Mrs. Gohil
Established: 1973
Open: Monday to Saturday 10am to 6pm
Stock: Ready made arts and crafts from India and Far East. Kalamkari silk and batik paintings. Hand painted gift items
Specialty: Indian arts and crafts
Prices: £5 to £500. Visa/Access/Master Card accepted

THE ROUND HOUSE GALLERY & SHOP
The Round House
Sheepscote Road and St. Vincent Road
Birmingham B16 8AE
Tel: 0121 608 5500
Established: 1996
Open: 9.30am to 5.30pm 7 days in winter. 9.30am to 7.30pm Thursday, Friday, Saturday and Sunday summer only
Crafts/Stock: Gallery: premium price one-off items made by members of Metis and touring exhibitions. Shop: craft and tourist items
Prices: Shop: up to £40
Note: Metis is a new organisation whose main aim is to promote British crafts. Located in Grade 1 listed building in centre of Birmingham. Also on site are craft worskhops, outdoor stage, restaurant, pub, stables

URBAN WOODEN PROJECTS
42A Church Road
Northfield
Birmingham
West Midlands B31 2LE
Tel: 0121 476 6569
Contact: Sue Kightley
Established: 1991

Open: 9am to 5.30pm Monday to Saturday. Closed Wednesday at 1pm
Stock: Wide variety of local crafts
Speciality: Dolls houses, furniture etc.
Prices: 99p to £550. Credit cards not accepted
Note: Spring and summer bedding plants and garden furniture. Children's playhouses

MARIAN WATSON DESIGNER JEWELLERY
2 Salop Street
Bishops Castle
Shropshire SY9 5BW
Tel: 01588 638864
Contact: Marian Watson
Established: 1991
Open: 10am to 5pm most days
Craftworkers on site: 1
Crafts: Silver and gold hallmarked designer jewellery. Caskets and trinket boxes as well as necklaces, bracelets, rings and ear-rings.
Prices: £30 to £1,000. Visa/Delta/ Mastercard/Eurocard accepted
Note: Individually designed and crafted jewellery strongly influenced by natural forms, often set with opals and moonstones, pearls and garnets, sometimes rubies and sapphires

THE ANDERSON GALLERY
96 High Street
Broadway
Worcestershire
WR12 7AJ
Tel: 01386 858086
Contact: Janet Edmondson
Established: 1990
Open: Daily 10am to 1pm, 2pm to 5pm.
Stock: High quality display of domestic pottery, studio ceramics, wood, metal, jewellery
Note: Contemporary crafts gallery in heart of Cotswolds. Monthly exhibitions. Commissions arranged

DAUB & WATTLE POTTERY AND GALLERY
Rear of 50 High Street (Windsor Street)
Bromsgrove
Worcestershire
B61 8EX
Tel: 01527 574004
Contact: Phill and Cath Cook
Established: 1988

Open: Monday to Saturday 9.30am to 5.30pm
Craftworkers on site: 3
Crafts/Stock: Studio pottery and ceramics
Prices: £3.50 to £250. Access/Visa accepted
Note: A working pottery where the production of ceramics in its various stages can be viewed on a daily basis

JINNEY RING CRAFT CENTRE
Hanbury
near **Bromsgrove**
Worcestershire
B60 4BU
Tel: 01527 821272
Fax: 01527 821869
Contact: Jenny or Richard Greatwood
Established: 1980
Open: All year Tuesday to Saturday 10.30am to 5pm. Sunday 11.30am to 5.30pm. Bank Holidays 10.30am to 5.30pm. Free Admission
Craftworkers on site: 13
Crafts: Repro furniture/antique restoration/french polishing. Milliner, children's clothes, violin maker, house signs, jewellery. Picture framer/photography, rocking horses, leather, decorative furniture, stained glass and pottery
Prices: £5 to £1,000. Visa/Access/Switch accepted
Note: Craft studios, restaurant serving teas, coffee, lunches. Gallery displaying crafts from all over Britain in monthly changing exhibitions. Also large gift shop. Limited disabled access

THE ROUND HOUSE GALLERY
38 High Street, Tutbury
Burton-on-Trent
Staffordshire
DE13 9LS
Tel: 01283 814964
Contact: Leah Evans
Established: 1988
Open: Monday to Saturday 11am to 5pm. Closed Wednesday and Sunday. Please telephone to check winter and Bank Holiday opening times
Stock: Domestic pottery, studio ceramics, glass
Prices: From under £5 to about £900
Note: Situated in historic village. A friendly atmosphere to view established and emerging ceramics and other crafts. Regular exhibitions

JAN BUNYAN POTTERY
4 Bridge Road
Butlers Marston
Warwickshire   CV35 0NE
Tel: 01926 641560
Contact: Jan Bunyan
Established: 1982
Craftworkers on site: Yes
Open: Usually 10.30am to 5.30pm Monday to Friday. Other times by arrangement. Advisable to telephone first
Crafts: Ceramics - tableware and decorative pieces
Prices: £10 to £100. Credit cards not accepted
Note: Commissions accepted. Professional member of the Craft Potters Association and member of the Gloucestershire Guild of Craftsmen

MONTPELLIER GALLERY
27 The Courtyard
Montpellier Street
Cheltenham
Gloucestershire GL50 1SR
Tel: 01242 515165
Fax: 01242 515165
Contact: Linda Burridge
Established: 1990
Open: Monday to Saturday 9.30am to 5.30pm throughout the year
Stock: Studio ceramics and glass, designer jewellery, paintings in oil and watercolour, printmaking, etchings, engravings and silkscreens
Note: An exhibition is held every other month either solo exhibitions or various artists brought together around a theme

SALLY GOYMER BASKETMAKER
37 Mendip Road
Cheltenham
Gloucestershire
GL52 5EB
Tel: 01242 510724
Contact: Sally Goymer
Established: 1979
Open: Most days, but please telephone for opening times
Craftworkers on site: 1
Craft: Traditional willow baskets. Commissions undertaken
Prices: £20 to £150. Credit cards not accepted
Note: Member of Basketmakers Association. Yeoman Member of the Worshipful Company of Basketmakers

J. M-J. POTTERY
Corner Workshop
189 Ashgate Road
Ashgate
Chesterfield
Derbyshire
S40 4AP
Tel: 01246 555461
Contact: Janine Mannion-Jones
Established: 1996
Open: 9am to 4.30pm Monday,
Wednesday and Friday. Closed Tuesday.
9am to 12.30pm Thursday. Please
telephone in advance of visit
Craftworkers on site: 1
Crafts/Stock: Domestic stoneware,
jewellery, some Raku and giftware
Prices: £2 to £200. Credit cards not
accepted

UNO GALLERY
5 South Street
Chesterfield
Derbyshire
S40 1QX
Tel: 01246 557145
Contact: Julie Benz
Established: 1991
Open: 10am to 5pm Monday to Saturday
Stock: Contemporary jewellery, metal,
glass
Prices: From £10 to £1000. Major credit
cards and Switch accepted
Note: Good selection of work from local
and national makers always in stock.
Crafts Council listed gallery

HARTS GUILD OF HANDICRAFT
The Guild, Sheep Street
Chipping Campden
Gloucestershire
GL55 6DS
Tel: 01386 841100
Contact: D.T. Hart
Established: 1908
Open: 9am to 5pm Monday to Friday.
Saturday 9 am to 12 noon
Craftworkers on site: 4 silversmiths
Crafts: Large scale, high quality silver
work from this long established and
unique firm - (not jewellery)
Prices: From under £30 to over £5000
upwards. Credit cards not accepted
Note: Featured on BBC Antiques Road
Show. The silversmiths may be viewed at
work and pieces can be viewed in new
exhibition area opened in 1996

FOREST STONECRAFT
Yew Tree Cottage
Bradley Hill, Soudley
near **Cinderford**
Gloucestershire
GL14 2UQ
Tel: 01594 824823
Fax: 01594 824317
Contact: Lesley Drew
Established: 1989
Open: 9am to 6pm daily
Craftworkers on site: 6
Crafts: Garden stoneware items
Prices: £1.99 to £450. Credit cards not
accepted
Note: Over 200 sculptured designs to
view and buy in 42 acres of landscaped
grounds. Gardens and water features
situated in the heart of the Forest of Dean
with guest house accommodation

BREWERY ARTS
Brewery Court
Cirencester
Gloucestershire
GL7 1JH
Tel: 01285 657181
Fax: 01285 644060
Contact: Tracey Burgoyne
Established: 1979
Open: 10am to 5pm
Craftworkers on site: 17
Crafts/stock: Ceramics, wire weaver,
letter press, ceramic sculptor, weaver,
furniture restorer, basket weaver, textile
workshop, restorer of antique rugs and
carpets, embroidery textile designer,
fashion embroiderer, jewellery, uphol-
sterer, stained glass artist
Prices: £5 to £250. Visa/Access/Switch
accepted
Note: Gloucestershire's unique multi-
disciplined arts and crafts centre. Crafts
Council Selected Craft Shop. Gallery, two
exhibition spaces, theatre, Egon Ronay
recommended coffee house, classes.
Limited disabled access.

**CHRIS ARMSTRONG COUNTRY
FURNITURE**
Paddock House
Clifford
Herefordshire
HR3 5HB
Tel: 01497 831561
Contact: Chris Armstrong
Established: 1990

Open: 9am to 5pm usually but advisable
to telephone
Craftworkers on site: 1
Crafts: Country furniture - ladderback
chairs, tables, turned ware
Prices: £20 to £500. Credit cards not
accepted
Note: Work is also on display at the
Rogues Gallery, Hay-on-Wye

CRICH POTTERY
Market Place
Crich
Derbyshire DE4 5DD
Tel: 01773 853171
Fax: 01773 857325
Contact: Diana Worthy
Established: 1975
Open: 10am to 6pm daily (telephone
beforehand if making a long journey on
Sundays)
Craftworkers on site: 4
Crafts: Stoneware ceramics handthrown
& decorated. Occasional prints/paintings
Prices: £3 to £200. Visa/MasterCard/
Eurocard
Note: No specific guided tours. Disabled
access downstairs and toilet

DENBY POTTERY VISITORS CENTRE
Derby Road
Denby
Derbyshire DE5 8NX
Tel: 01773 740799
Fax: 01773 570211
Contact: Tours Reception
Established: 1993
Open: 9am to 5pm Monday to Saturday.
Tuesday 9.30am to 5pm. Sunday 10am to
5pm
Crafts: Denby stoneware, tableware and
cookware
Price: £3 to £160. Switch/Visa/Delta/
Mastercard accepted
Note: Restaurant, children's play area,
florist, Dartington crystal factory shop,
craftroom, second quality Denby pottery,
cookery emporium, demonstration
kitchen, guided tours. Disabled access

NIGEL GRIFFITHS
The Old Cheese Factory
Grangemill
Derby
Derbyshire DE4 4HG
Tel: 01629 650720
Contact: Nigel Griffiths

Established: 1946
Open: 8.30am to 5pm
Craftworkers on site: 4
Crafts: Oak furniture
Prices: £250 to £4,500. Credit cards not
accepted

OMBERSLEY GALLERY
Church Terrace, Ombersley
near **Droitwich Spa**
Worcestershire WR9 0DP
Tel: 01905 620655
Fax: 01905 620655
Contact: Carole Pimm
Established: 1981
Open: 10am to 5pm Tuesday to Saturday
Stock: High quality ceramics, pottery,
wood, glass, jewellery plus original
paintings and etchings
Prices: £30 to £5,000. Visa/Access/
Switch accepted
Note: Egon Ronay selected restaurant and
tearoom. Home cooking – seasonal menu

MORIAD CRAFTS
Fletcher's Garden Centre
Bridge Farm, Stone Road
Eccleshall
Staffordshire ST21 6JY
Tel: 01785 850191
Contact: Moriad York-Batten
Established: 1990
Open: 11am to 4pm Wednesday,
Thursday, Friday. 11am to 5pm Saturday
& Sunday. Closed Monday & Tuesday
Stock: Pottery, textiles, cards, jewellery,
gifts and craft work, cross stitch kits, hand-
made soft toys, brass ware, leather goods
Speciality: Cushion panels
Prices: £1 to £25. No credit cards
Note: Situated in large garden centre.
Tearooms (lunches also), birds of prey

THE KOUYZER POTTERY
Church Street
Edwinstowe
Nottinghamshire NG21 9QA
Tel: 01623 825900
Fax: 01623 825900
Contact: Peter Kouyzer
Established: 1992
Open: Daily 10am to 6pm
Craftworkers on site: 4
Craft: Quality stoneware pottery
Prices: 95p to £120. Credit cards accepted
Note: Demonstrations held. Craft shop &
gallery. Robin Hood oak 400 yards away

ANNIE'S ATTIC
Studio 10, Craft Complex
Oxford Road
Hay-on-Wye
Herefordshire HR3 5DG
Tel: 01497 821226
Contact: Annie Lewis
Established: 1991
Open: Daily 10am to 4pm with seasonal
variations
Crafts: Salt dough decorations handmade
by the owner on the premises
Prices: £1 to £15. Credit cards not
accepted
Note: Situated in Hay-on-Wye - the town
of books and crafts

BROOK STREET POTTERY
Brook Street
Hay-on-Wye
Herefordshire HR3 5BQ
Tel: 01497 821026
Contact: Simon or Sara Hulbert
Craftwokers on site: 2
Established: 1994
Open: Monday to Saturday 10.30am to
5pm. Sunday 12 noon to 4.30pm. Closed
Tuesdays
Prices: From £10 to £500. All major
credit cards accepted plus Switch & Delta
Crafts/stock: Domestic pottery, contem-
porary tableware, garden terracotta as
well as metal, textiles, baskets
Note: A working pottery selling their own
ware and other crafts

MERIDIAN
13 High Town
Hay-on-Wye
Herefordshire HR3 5AE
Tel: 01497 821633
Contact: Tony Britnell
Established: 1987
Open: Wednesday to Saturday 10am to
5pm. Sunday 2pm to 5pm
Stock: Studio ceramics, jewellery.
Contemporary fine and applied art. Work
of national and international renown
Prices: £30 to £10,000. Switch/Delta

THE ROGUES GALLERY
2 Broad Street
Hay-on-Wye
Herefordshire
HR3 5DB
Tel: 01497 821677
Contact: Chris Armstrong

Established: 1995
Open: 11am to 5pm Thursday, Friday,
Saturday and Monday, Tuesday,
Wednesday in holiday periods
Stock: Furniture, paintings and prints,
hats, jumpers, jackets, pottery, cards,
coppice products, turned bowls. All
locally made
Prices: £1 to £500. Credit cards not
accepted
Note: Exhibitions from time to time
featuring particular crafts

DORE MILL
Abbeydore
Hereford
Herefordshire
HR2 0AA
Tel: 01981 240464
Contact: Lorna and Geoff Hollom
Established: 1988
Open: Most days, but please telephone
for opening times
Craftworkers on site: 2
Crafts: Country chairs and stools,
cushions and pictures in woven and
felted wool. 'Woolly' greetings cards
Prices: £1 to £240. Credit cards not
accepted
Note: Members of Hereford Society of
Craftsmen. 12th century Dore Abbey and
Abbeydore Court Gardens nearby

JOHN MCKELLAR DESIGNER
JEWELLERY
23 Church Street
Hereford
Herefordshire
HR1 2LR
Tel: 01432 354460
Contact: John McKellar
Established: 1984
Open: 9.30am to 5.30pm Monday to
Saturday
Craftworkers on site: Yes
Craft/Stock: Jewellery in all materials -
precious and non-precious metals.
Ceramics, plastics, textiles, etc.
Speciality: Jewellery exhibitions every
month
Prices: £10 to £1500. Visa/Mastercard/
Access/American Express/Diners Club
accepted
Note: Jewellery workshop on premises.
Commissions undertaken. Crafts Council
listed. Disabled access difficult but not
impossible – two steps up from street level

GALLERY CRAFTS
4 Bank Gallery
High Street
Kenilworth
Warwickshire
CV8 1LY
Tel: 01926 851720
Contact: John Flower
Established: 1993
Open: 10am to 5pm Tuesday to Saturday
Stock: Hand made, high quality British
crafts - sculpture, ceramics, turned wood,
prints, original paintings and picture
framing
Speciality: Small sculptured objects by
local artists
Prices: 40p to £700. Visa/Mastercard/
Amex accepted
Note: Free parking available. Cafe
adjacent. Gardens, Abbey fields, old part
of Kenilworth and castle nearby

ENGLISH OAK FURNITURE
8/10 Headbrook
Kington
Herefordshire
HR5 3DZ
Tel: 01544 230208
Contact: Terence William Clegg
Established: 1977
Open: Monday to Saturday 8am to 5pm.
Sunday 11am to 5pm
Craftworkers on site: 3
Crafts: Traditional English furniture
mainly oak and walnut
Prices: £100 to £2,000. Credit cards not
accepted
Note: Parking available

HIBISCUS
109 Regent Street
Leamington Spa
Warwickshire
CV32 4NV
Tel: 01926 337716
Fax: 01926 337716
Contact: Liz Silverston
Established: 1993
Open: 10am to 5.30pm Monday to
Saturday
Stock: Precious jewellery (platinum, 18ct
gold and diamonds). Non-precious
jewellery, ceramics, glass – emphasis on
colour and wackiness
Prices: £5 to £1500. Diners/Visa/Amex/
MasterCard
Note: Three exhibitions held each year

COTSWOLD WOOLLEN WEAVERS
Filkins

near LECHLADE
Gloucestershire
GL7 3JJ
Tel: 01367 860491
Fax: 01367 860661
Email: wool.weavers@dial.pipex.com
Contact: Richard Martin
Established: 1982
**Open: 10am to 6pm Monday to
Saturday. 2pm to 6pm Sunday**
Craftworkers on site: 10 plus
**Crafts: Spinning and weaving
woollen cloth. Wide range garments,
knit wear, rugs and accessories – all
woven on the premises**
**Prices: £1 to £200. Visa/Mastercard
accepted**
**Note: Traditional working mill
housed in splendid 18th century
buildings. Permanent exhibition
areas, coffee shop and picnic area.
No admission charge. Coach parties
must book in advance**

FILKINS GALLERY AND STUDIO
Cross Tree Yard
Filkins
near **Lechlade**
Gloucestershire
GL7 3JJ
Tel: 01367 850385
Contact: Sallie Seymour and Wendy
Robinson
Established: 1993
Open: 10.30am to 4.30pm October to
March. 10.30am to 5pm April to
September
Craftworkers on site: 2
Crafts/Stock: Paintings, ceramics, wood
and various textiles that are all locally
made
Prices: £1.25 to £300. Visa/Access/
Mastercard accepted
Note: Other workshops close by.
Cotswold Woollen Weavers with teashop
in same yard. Exhibitions are held
spotlighting a specific theme, artist or
craft every two months. Disabled access
is possible

OLD BELL POTTERY
High Street
Lechlade
Gloucestershire
GL7 3AD
Tel: 01367 252608
Contact: Keith Broley
Established: 1974
Open: 9.30am to dusk
Craftworkers on site: 1
Crafts: Garden terracotta
Prices: £1 to £100. Visa/MasterCard
accepted
Note: Boxwood topiary trees for sale.
Humorous garden and canary aviary

COLLECTION GALLERY
The Southend
Ledbury
Herefordshire
HR8 2EY
Tel: 01531 634641
Contact: Clare Stacpoole
Established: 1980
Open: 10am to 5.30pm daily, some
Sundays – please ring
Stock: Ceramics, jewellery, glass, wood
and textiles
Speciality: Ceramics
Prices: £10 to £1,000. Mastercard/Amex/
Access/Visa accepted
Note: Regular exhibitions, talks and trips.
Contact gallery for exhibition programme

EASTNOR POTTERY
Clock Cottage
Home Farm
Eastnor
near **Ledbury**
Herefordshire
HR8 1RD
Tel: 01531 633255
Fax: 01531 633255
Contact: Jon Williams and Sarah Monk
Established: 1994
Open: 10am to 5pm Sundays April to end
September. Other times by arrangement
Crafts: One-off and small production
domestic ware and studio ceramics
Speciality: Breakfastware by Sarah Monk.
Studio ceramics by Jon Williams
Prices: £6.50 to £300. Credit cards not
accepted
Note: Eastnor Castle and deer park and
garden centre nearby. Other Eastnor crafts
include textile designers, artist and
blacksmith

HOMEND POTTERY
205 The Homend
Ledbury
Herefordshire
HR8 1BS
Tel: 01531 634571
Contact: Mark and Caroline Owen-
Thomas
Established: 1981
Open: 10.30am to 5pm Tuesday to
Saturday. Closed Wednesdays pm
Craftworkers on site: 2
Crafts: Hand-thrown and decorated
creamware pottery
Prices: £5 to £100
Note: This is a full time working pottery -
and work can often be seen in the making
stages

LEDBURY CRAFT CENTRE
1 High Street
Ledbury
Herefordshire
HR8 1DS
Tel: 01531 634566
Contact: Peter and Carol Preston
Established: 1985
Open: 9.15am to 5.30pm
Stock: Candles, pottery, pot pourri, corn
dollies, jewellery, books, cards, toys
and gifts
Speciality: Source books for the craft-
worker and teachers
Prices: Pence to pounds. Access/Visa
accepted

**CUCKOO'S NEST CRAFT SHOP &
POTTERY**
3 Brookbridge Court
Melton Road
Syston
Leicester
Leicestershire
LE7 2JT
Tel: 0116 260 6286
Contact: Mick and Kate Glover
Established: 1991
Open: 9.30am to 5pm daily. Closed on
Sundays and Bank Holidays
Crafts/stock: Original paintings and a
variety of crafts and gifts as well as
owners own pottery that is produced on
the premises
Prices: £1 to £100. Visa/Mastercard/
Switch accepted
Note: Teddies' hospital. Coffee shop next
door

NUMBER 34
34 Tamworth Street
Lichfield
Staffordshire
WS13 6JJ
Contact: Jennifer C. Hobbs
Established: 1992
Stock: British contemporary pottery,
picture frames (made in studio), screen
prints, etchings, lino prints, woodcuts
Prices: £5 to £300. Access/Visa
accepted
Note: Cleaning, conservation and
restoration of frames and pictures
undertaken

MOHAIR, SILK & CRAFT CENTRE
Blakemore Farms
Little London
Longhope
Gloucestershire
GL17 0PH
Tel: 01452 830630
Contact: Mrs. Y. Williams
Established: 1993
Open: 10.30am to 5pm. Closed
Mondays and Tuesdays except Bank
Holidays
Stock: Mohair knitwear and garments
produced from angora goats bred on the
farm. Local crafts also on sale
Speciality: Mohair products and silk
blouses
Prices: 50p to £150. Access/Visa/
Mastercard accepted
Note: Farm walks, coffee shop, disabled
facilities. Free entry and car park.
Coaches and groups by appointment.
Animal (goat) rescue centre. Donation
boxes displayed

MARIA DALMAR – LACE
37 Atherstone Road
Loughborough
Leicestershire
LE11 2SH
Tel: 01509 266302
Contact: Maria Dalmar
Established: 1972
Open: Telephone for appointment
Crafts: All forms of lace from doileys to
curtains
Speciality: Lace appliqué cushions
Prices: £3.50 to £400. Credit cards not
accepted
Note: Talks and demonstrations given.
Crafts Council listed

WYMESWOLD COUNTRY FURNITURE
17 Far Street, Wymeswold
Loughborough
Leicestershire LE12 6TZ
Tel: 01509 880309
Fax: 01509 880309
Contact: Bill and Jenny McBean
Established: 1984
Open: 9.30am to 5.30pm Monday to
Saturday. 2pm to 5pm Sunday
Craftworkers on site: 2
Crafts/stock: Furniture and antique
restoration, knitwear, iron work,
jewellery, paintings, pottery, needlework
and clocks
Prices: £1 to £1,000. Visa/Mastercard
accepted
Note: The 6 showrooms and 3 workshops
are housed in converted Georgian stables.
Furniture is delivered in a restored 1936
Bedford van.

CROSSING CRAFTS
The Crossing
Ashford Bowdler
near **Ludlow**
Shropshire
SY8 4DJ
Tel: 01584 831532
Fax: 01584 831532
Contact: Louise Herriott
Established: 1993
Open: Most days but please phone first
Craftworkers on site: 2
Crafts: Quilted cushions and bags, cat
doorstops and draught excluders
Prices: £3.50 to £20. Credit cards not
accepted

WOODSTOCK HOUSE CRAFT CENTRE
Woodstock House
Brimfield
near **Ludlow**
Shropshire
SY8 4NY
Tel: 01587 711445
Contact: Wendy Rulton
Established: 1983
Open: 9am to 5.30pm
Craftworks on site: 6 part time
Craft/stock: Jackets, textural patchwork
jackets in corduroy and silk, appliqué
Prices: £80 to £180
Note: Two large fabric wall hangings in
local churches. Large fabric and
haberdashery shop. Craft holidays
organised. No disabled access

ANDREW J. PYKE CRAFTSMAN IN WOOD
Grove Farm
Brockwier Lane, Hewelsfield
Lydney
Gloucestershire
GL15 6UU
Tel: 01594 530924
Contact: Andrew and Susan Pyke
Established: 1978
Open: Any reasonable hour but please telephone first
Craftworkers on site: 1
Crafts: Furniture, antique repairs, high class joinery, kitchen units, church furniture
Prices: All work is designed and made to commission
Note: A wide range of temperate hardwoods are seasoned on the premises which can be viewed. Photographic record of all work. Crafts Council listed

ANNIE DYSON DESIGNS
Flat 1, Burford House
32 Worcester Road
Malvern
Worcestershire
WR14 4QW
Tel: 01684 563556
Contact: Annie Dyson
Open: 9am to 5pm
Crafts: Hand-made greetings cards in watercolours. Very bright, unique and individual. Also watercolour paintings/illustrations
Prices: £2 to £70

HIBERNIAN VIOLINS
24 Players Avenue
Malvern
Worcestershire
WR14 1DU
Tel: 01684 562947
Contact: Padraig o Dubhlaoidh
Established: 1988
Open: 9am to 5pm Monday to Friday or by appointment
Craftworkers on site: 1
Crafts: Hand made musical instruments of the violin family. Antique instruments, bows etc.
Prices: £200 to £7,000. Credit cards notaccepted
Note: Instruments available for trial. Wheelchair accessible. Malvern Hills nearby and Morgan car factory

EAST CARLTON COUNTRYSIDE PARK
East Carlton
near **Market Harborough**
Leicestershire LE16 8YF
Tel: 01536 770977
Fax: 01536 770661
Contact: Chris Smith
Established: 1989
Open: 9.30am to 4.45pm summer. 9.30am to 3.45pm winter
Craftworkers on site: 6
Crafts/stock: Glass figures, dried flowers, jewellery, joinery, hand painted ceramics, woollens
Prices: £2 to £500. No credit cards
Note: Cafe, 100 acres of countryside park, steel heritage centre. Wildlife interpretation centre. No disabled access

QUORN POTTERY
46-48 Scotland Road, Little Bowden
Market Harborough
Leicestershire LE16 8AX
Tel: 01858 431537
Fax: 01858 431537
Contact: J.W. Brookes
Established: 1977
Open: 9am to 5pm weekdays. 10am to 4pm Saturday
Craftworkers on site: 3
Crafts: Slip cast novelty teapots, clocks, mugs, cruets
Prices: £1.25 to £39.95. No credit cards

THE FRANK HAYNES GALLERY
50 Station Road
Great Bowden
Market Harborough
Leicestershire LE16 7HN
Tel: 01858 464862
Contact: Frank Haynes
Established: 1987
Open: Thursday to Sunday 10am to 5pm all year round
Stock: Pottery by Frank Haynes and other members of the Midland Potters Association and the CPA etc. Varied work from 15 people, currently
Speciality: Bowls decorated with figures
Prices: £3 to £175. Major credit cards accepted
Note: Art gallery with paintings from Leicestershire and Northants artists. Also cards, picture framing and clay. Large group (bus) visits must be notified two weeks in advance please. Disabled access

CAUDWELLS MILL CRAFT CENTRE
Rowsley
Matlock
Derbyshire
DE4 2EB
Tel: 01629 733185
Contact: Mr. R.D. Priestley
Established: 1987
Open: 10am to 6pm 1 April to 31 October.
10am to 4.30pm. 1 November to 31 March.
Weekends only January, February
Craftworkers on site: 5
Crafts/stock: Glassblowing, pottery,
wood turning, artist and furniture restorer.
Both the craft shop and gallery sell work
by local, national and international
makers
Prices: £1 to £500. Visa/Mastercard
accepted
Note: Tearooms, working water-powered
flour mill, nature trail, sited in Victorian
mill buildings. Delightful countryside in
the Peak National Park. Disabled access

HOTHOUSE
7 Lumsdale Mill, Lower Lumsdale
Matlock
Derbyshire
DE4 5EX
Tel: 01629 580821
Fax: 01629 580821
Contact: Anthony Wassell
Established: 1987
Open: Most days but please 'phone first
Craftworkers on site: 2/3
Crafts: Free blown coloured contempo-
rary studio glass
Prices: £1 to £450. Visa/Mastercard
accepted
Note: Visitors are able to watch from our
purpose built viewing area when
glassmaking is in progress. No charge
during daytime hours. Large parties by
arrangement. Disabled access at rear

MELBOURNE HALL
Melbourne
Derbyshire
DE73 1EN
Tel: 01332 862502
Contact: Mrs G. Weston (Hall Curator)
Established: 1980s
Open: February to December daily.
Closed Mondays except Bank Holidays
Craftworkers on site: 10
Crafts/stock: Gifts, dried flowers and oils,
furniture restoration, wood turner, knit-

wear, celebration cakes, photographer,
reupholstery, picture framer and gallery
Prices: 50p to £200. Enquire direct to
craft units for credit card acceptance
Note: Tearooms, historic garden (only
open at certain times of year). Disabled
access

RUARDEAN GARDEN POTTERY
Ruardean
Mitcheldean
Gloucestershire
GL17 9TP
Tel: 01594 543577
Contact: John Huggins
Established: 1979
Open: 9.30am to 5.30pm Monday to
Saturday. 1pm to 5pm Sunday (April to
September)
Craftworkers on site: 3
Crafts: High quality, hand made, frost
proof terracotta flower pots and garden
ornaments
Prices: £3 to £950. Credit cards not
accepted
Note: Pots can be seen being made

CLEEHILLS CRAFTS
Stanton Long
Much Wenlock
Shropshire TF13 6LH
Tel: 01746 712677
Contact: Carol Colley
Established: 1989
Open: 10am till dusk all year
Craftworkers on site: 4
Crafts: Pottery (terracotta) and black wool
products. Modelled floral terracotta ware
Prices: 50p to £50
Note: Farm walk with picnic areas. Black
rare breeds. Caravan site

DAVE TIGWELL BLACKSMITH
Ladywood
Whips Lane, Watledge
Nailsworth
Gloucestershire
GL6 0BB
Tel: 01453 832954
Contact: Dave Tigwell
Established: 1988
Open: Most days but please 'phone first
Craftworkers on site: 1
Crafts: Curtain rails, candle sconces, door
furniture etc. Iron/brass
Note: All kinds of metalwork undertaken,
specialising in Ernest Gimson's designs

JOY'S CARDS & CRAFTS
18A High Street, May Bank
Newcastle-under-Lyme
Staffordshire ST5 0JB
Tel: 01782 713297
Contact: Joy Tune
Established: 1990
Open: 9.30am to 5pm daily. Closed all day Thursday and Sunday
Stock: Wide range of locally made crafts, material-made goods: baskets, cushions, baby items very reasonably priced; dried flower arrangements. Range of gift items. Cross-stitch kits
Prices: 99p to £50. No credit cards

COWDY GALLERY
31 Culver Street
Newent
Gloucestershire GL18 1DB
Tel: 01531 821173
Fax: 01531 821173
Contact: Harry Cowdy
Established: 1989
Open: Tuesday to Friday 10am to 12.30pm and 1.30pm to 5pm. Saturday 10am to 12.30pm. Hours vary
Stock: Contemporary glass by designer/makers and artists
Prices: £5 to £2,500. Visa/Mastercard accepted
Note: Annual programme of exhibitions

NEWENT SILVER AND GOLD
15 Bread Street
Newent
Gloucestershire
GL18 1AQ
Tel: 01531 822055
Contact: Mr. K. Vowles
Established: 1987
Open: 9am to 1pm and 2pm to 5pm Monday to Saturday
Craftworkers on site: 1
Crafts: Gold and silver jewellery and giftware
Prices: £1 to £2,000. Visa/Mastercard accepted
Note: Jewellery-making demonstrations

CASTLE MUSEUM CRAFT SHOP
Castle Museum
Nottingham NG1 6EL
Tel: 0115 9153662
Contact: Rachel Dewsbury
Established: 1986
Open: Monday to Sunday 10am to 5pm

Stock: Domestic pottery, studio ceramics, metal, jewellery, toys and cards
Prices: From under £10 up to £200. Credit cards/Switch accepted
Note: 17th century mansion with landscaped grounds. Collection of fine & applied art. Regularly changing craft display

CRUSOE DESIGNS
4th Floor, 4 Castle Boulevard
Nottingham
Nottinghamshire NG7 1FB
Tel: 0115 941 7475
Contact: Trevor Naylor
Established: 1995
Open: 10am to 4pm usually Monday to Saturday but worth ringing first
Craftworkers on site: 1
Crafts: Furniture and mirrors made from reclaimed pine
Prices: Under £40 to just under £500
Note: Complete interior design service offered. Five minutes walk from Nottingham Castle. Back door opposite The Trip to Jerusalem (oldest pub in England)

FOCUS GALLERY
108 Derby Road
Nottingham
Tel: 0115 953 7575
Contact: Margaret James
Established: 1971
Open: Monday to Saturday 9.30am to 5.30pm (10am Mondays)
Stock: Studio pottery, glass, jewellery, woodcarvings, candles and greetings cards as well as original prints
Prices: From under £10 up to £500. All major credit cards accepted
Note: Work exhibited by approx. 40 potters, 10 glassblowers, 20 jewellers, 25 printmakers, 12 craftsmen in wood or metal and 20 greetings card designers

GRANGE FARM POTTERY
Grange Farm
Plungar
Nottingham
Nottinghamshire
NG13 0JJ
Tel: 01949 860630
Contact: Elaine Pell
Established: 1990
Open: 11am to 6pm Saturday and Sunday. Closed Christmas to Easter
Craftworkers on site: 1
Crafts: Stoneware ceramics produced on

premises. Large garden pots, frostproof
not terracotta
Prices: 50p to £65. Credit cards not
accepted
Note: Workshop is open at other times if
the owner is there; please phone first

PATCHINGS FARM ART CENTRE
Oxton Road
Calverton
Nottingham
Nottinghamshire NG14 6NU
Tel: 0115 965 3479
Fax: 0115 965 5308
Contact: Liz Wood
Established: 1988
Open: Picture framing, art materials and
craft shop open 9am to 6pm daily.
Galleries and restaurant 9am to 10.30pm
daily
Crafts/stock: Pottery, pictures - prints and
original works of art - cards, candles,
jewellery
Prices: £3 to £5,000. Visa/Access
accepted
Note: Monthly exhibitions, mainly two
dimensional plus textiles. Restaurant,
gardens, art school, National Art and
Designer Craftsman event early each
June. Disabled access. Accommodation
available on site. Art demonstrations
available for groups by prior booking

BOSWORTH CRAFTS
23 Main Street, Market Bosworth
Nuneaton
Warwickshire CV13 0JN
Tel: 01455 292061
Contact: R. Thorley
Established: 1981
Open: 9am to 5pm Monday to Saturday
Craftworkers on site: 1
Crafts: All types of leather goods. Pottery,
figures, cottages, plates
Speciality: Hand carved leather
Prices: 85p to £80. Visa/MasterCard/
Diners Club/American Express accepted
Note: Demonstrations of leather carving

DRAGONPOT CRAFTS
3 Brooksby Drive
Oadby
Leicestershire LE2 5AA
Tel: 0116 271 1337
Contact: Val Chivers
Established: 1992
Open: 10am to 5pm

Crafts/stock: Local crafts. Pottery items
made on premises
Speciality: Humorous dragons
Prices: £5 to £50 dragons. £50 upwards
for 10" high figures. Visa/Access accepted

RUFFORD GALLERY
Rufford Country Park
near **Ollerton**
Nottinghamshire NG22 9DF
Tel: 01623 822944 **Fax:** 01623 824702
Contact: Peter Dnorok
Established: 1981
Open: January to February 11am to 4pm
March to December 10.30am to 5pm
Stock: Contemporary ceramics
Prices: £10 to £500. All credit cards
accepted
Note: Craft shop, souvenir shop, buttery
restaurant, coach house, coffee shop,
sculpture collection, plant and herb
garden, cistercian abbey, country park
and lake, woodland walks

TICKMORE POTTERY
Treflach Road, Trefonen
Oswestry
Shropshire SY10 9DZ
Tel: 01691 661842
Fax: 01691 661842
Contact: Felicity Cripps
Established: 1979
Open: Most days except some weekends
Craftworkers on site: 1
Crafts: Hand thrown stoneware pottery.
Mostly domestic ware, some decorative
Prices: £2.50 to £100. Credit cards not
accepted
Note: The pottery is known for its range
of 'sheep' decorated ware. Visitors are
welcome to watch processes and view

THE CRAFT STUDIOS
43 Main Street, Woodnewton
Oundle
Northamptonshire PE8 5EB
Tel and Fax: 01780 470866
Contact: Rob Bibby
Established: 1989
Open: Weekdays. Please telephone for
weekend opening
Craftworkers on site: 3
Crafts: Maiolica pottery, woodcarving
and puppet making (marionettes)
Prices: £2.50 to £1,000. Access/Visa/
MasterCard accepted
Note: Talks given. Disabled access

PAINSWICK WOODCRAFTS
3 New Street
Painswick
Gloucestershire
GL6 6XH
Tel: 01452 814195
Contact: Dennis and Desiree French
Established: 1992
Open: Tuesday to Saturday 9.30am to
5pm except for 2 weeks in January. Open
Sunday afternoons in season. Please
phone for full details
Crafts/stock: Speciality woodware. Also
local pottery, prints, calligraphy, silk tie-
dyed scarves, book binding, small
furniture
Speciality: Domestic woodware by
Dennis French
Prices: 30p to £270. Mastercard/Visa
accepted
Note: Very attractive Cotswold village,
ample tea/coffee shops, pub lunches.
Famous yew trees of St. Mary's Church

OLD CHAPEL GALLERY
East Street
Pembridge
near Leominster
Herefordshire
HR6 9HB
Tel: 01544 388842
Contact: Yasmin Strube
Established: 1988
Open: Monday to Saturday 10am to
5.30pm. Sunday 2pm to 5.30pm
Stock: High quality domestic pottery,
studio ceramics, wood, contemporary
furniture, metal, iron work, jewellery,
textiles, knitwear, glass, toys, garden
sculpture also original paintings & prints
Prices: From £5 to around £1000. Major
credit cards and Switch
Note: Permanent display of local art and
craft, also country furniture. Exhibitions
held throughout the year

LONGDALE CRAFT CENTRE &
MUSEUM
Longdale Lane
Ravenshead
Nottinghamshire NG15 9AH
Tel: 01623 794858
Fax: 01623 794858
Established: 1971
Open: Daily
Craftworkers on site: 30
Crafts/stock: Sculpture, silversmith,

stained glass, pottery. All restoration. Art
materials and equipment
Prices: £50 upwards. Credit cards not
accepted
Note: Shop, gallery, craft village museum
and restaurant on site. Parking available

CHRIS ASTON CERAMICS
The Pottery
4 High Street, Elkesley
near **Retford**
Nottinghamshire DN22 8AJ
Tel: 01777 838391
Fax: 01777 838391
Contact: Chris Aston
Established: 1970
Open: All week
Craftworkers on site: 2
Crafts: Hand thrown stoneware, lustre
and screenprinted enamel items
Prices: £4.50 to £150. Credit cards not
accepted
Note: Visitors are able to see work in
progress. Group formal talks and
demonstrations can be booked

WOBAGE FARM WORKSHOPS
Upton Bishop
Ross-on-Wye
Herefordshire HR9 7QP
Tel: 01989 780 495/233
Contact: Mick Casson
Established: 1977
Open: Saturday and Sunday 10am to
5pm. Other times by appointment
Craftworkers on site: 9
Crafts/stock: Highly individual and
varied stoneware pottery, domestic and
gardenware, thrown and handbuilt.
Bespoke furniture and carved furniture.
Precious jewellery
Prices: £2 to £2,000. Credit cards not
accepted
Note: Commissions taken. Permanent
exhibition in spacious showroom.
Principal outlet for Mick and Sheila
Casson's salt-glazed pottery. Annual
Christmas exhibition. Close to pubs and
restaurants, less than 90 minutes from
Birmingham, Cardiff and Oxford

BRINKLOW POTTERY
11 The Crescent, Brinklow
near **Rugby**
Warwickshire
CV23 0LG
Tel: 01788 832210

Contact: George and Diane Lindsay
Established: 1985
Open: Gallery: Tuesday to Saturday
10am to 6pm, closed Sunday and
Monday. Workshop only by arrangement
Craftworkers on site: 2
Crafts: Functional and decorative
stoneware. Commemorative ware.
Tableware, kitchenware, house names,
dragons and other sculptural items.
Fragrance and aromatherapy oils
Prices: £5 to £200. Cheques accepted
with guarantee card. No credit cards
Note: Near 11th century church in att-
ractive village location, pubs, eating places

NEVILLE AND LAWRENCE NEAL

22 High Street
Stockton
near **Rugby**
Warwickshire
CV23 8JZ
Tel: 01926 813702
Contact: Neville and Lawrence Neal
Established: 1958
Open: 9am to 5pm daily. Weekends by
appointment
Craftworkers on site: 2
Crafts: Spindle and ladderback chairs
with rush seats (dining chairs, armchairs
and rocking chairs)
Prices: £75 to £275. Credit cards not
accepted
Note: Chairmaking demonstrations. A
village craft founded by Ernest Gimson at
the turn of the century

ST. JULIAN'S CRAFT CENTRE

Wyle Cop/High Street
Shrewsbury
Shropshire
SY1 1UH
Tel: 01743 353516
Contact: Andrew S.N. Wright
Established: 1980
Open: 10am to 5pm Monday to Saturday.
Closed Thursdays exc. school holidays.
10am to 4pm January and February
Crafts/stock: Wildlife products, soft toys
and bird boxes, hardwood toys,
kitchenware, slatework clocks and house
signs, letterpress printing, calligraphy,
shoes, craft supplies, beads, pine
furniture, wooden letters and names,
dried flowers, paintings and frames, pottery
Prices: 25p to £100. Some studios accept
credit cards

TORQUIL POTTERY

81 High Street
Henley-in-Arden
Solihull
West Midlands B95 5AT
Tel: 01564 792174
Contact: Reg Moon
Established: 1960
Craftworkers on site: 1
Crafts: Stoneware and porcelain pottery.
Fine art and crafts
Prices: £5 to £200
Note: Exhibitions in Gallery upstairs of a
wide range of fine art and craft

DESIGNS IN WOOD

Amerton Farm
Stowe-by-Chartley
near **Stafford**
Staffordshire ST18 0LA
Tel: 01889 271193 **Fax:** 01889 271193
Contact: David Hanlon
Established: 1990
Open: 10.30am to 5pm daily
Craftworkers on site: 2
Crafts: Woodturning and bespoke
furniture. Pottery, fabrics, dried flowers
Prices: 95p to £1,500. Visa/Mastercard
accepted
Note: Glass sculptor. Woodturning
lessons by appointment. Garden centre,
bakery, tearoom, farm animals and shop,
wildlife centre. Large free car park

MIDLAND CRAFTS

Cromwell House, Wolseley Bridge
near **Stafford**
Staffordshire ST17 0XS
Tel: 01889 882544
Fax: 01889 882544
Contact: G.J. Foulkes
Established: 1981
Open: Weekdays 10am to 5pm. Saturday
and Sunday 1.30pm to 5pm
Stock: Range of onyx goods. Lighting,
tables and giftware. Probably largest stock
in UK of giftware in woodstone and fossil
coral. Gemstones, mineral specimens and
associated jewellery
Speciality: Stone crafts
Prices: £1 to £250. All credit cards
accepted
Note: Demonstrations for parties by
appointment. Set in superb part of
Cannock Chase. Large car park, picnic
area, garden centre, cafe and other craft
shops close by

SHIRE HALL GALLERY
Market Square
Stafford
Staffordshire ST16 2LD
Tel: 01785 278345
Fax: 01785 278327
Contact: Mrs. Hilary Foxley
Established: 1993
Open: Monday to Saturday 10am to 5pm
Stock: High quality domestic pottery,
studio ceramics, glass, wood, jewellery,
textiles and toys
Prices: From under £20 to over £200.
Credit cards accepted
Note: Housed in magnificent 18th
century building. High quality British
crafts from new and established makers

CITY MUSEUM AND ART GALLERY
Bethesda Street
Hanley
Stoke-on-Trent
Staffordshire ST1 3DW
Tel: 01782 202173
Fax: 01782 205033
Contact: Mr. Alan Townsend
Established: 1980
Open: Monday to Saturday 10am to 5pm.
Sunday 2pm to 5pm (Museum opening
times)
Stock: Ceramics and jewellery
Speciality: Ceramic books
Note: Museum contains largest collection
of English ceramics in the world.
Ceramics gallery has more than 5000
valuable pieces on display
Note: Museum shop caters for all tastes, a
ceramic books booklist is available

PETER GOSLING STUDIO
370A Ruxley Road
Bucknall
Stoke-on-Trent
Staffordshire
ST2 9BA
Tel: 01782 219106
Contact: Mr. or Mrs. P. Gosling
Established: 1975
Open: Monday to Thursday 2pm to 6pm
Crafts/stock: Pottery, watercolours, oil
paintings and exclusive prints
Speciality: China painting
Prices: 50p to £200. Credit cards not
accepted
Note: Bone china fancies made on the
premises and demonstrations held if
required

MCCUBBINS
St. Briavels
Gloucestershire
GL15 6TQ
Tel: 01594 530297
Contact: Gill McCubbin
Established: 1979
Open: Easter Saturday to Christmas -
Thursday, Friday, Saturday, Sunday,
Monday 10am to 1pm and 2pm to 5pm.
December 30th to Good Friday -
Saturday and Sunday only 10am to 1pm
and 2pm to 5pm
Craftworkers on site: 3
Crafts: Stoneware domestic pottery/
contemporary silver jewellery. Many
crafts including cats and clocks in glass,
wood and metal
Prices: £1.25 to £800. All major credit
cards accepted
Note: Located in Forest of Dean
overlooking Wye Valley. Beautiful walks,
moated castle, YHA, pubs, teashop.

FORGING AHEAD
Field House Farm
Belbroughton
Stourbridge
Worcestershire
DY9 9SS
Tel: 01562 730003
Fax: 01562 730003
Contact: Paul Margetts
Established: 1989
Open: Usually 9am to 5pm Monday to
Friday by appointment
Craftworkers on site: 1
Crafts: Forged metalwork - gates,
candlesticks, clocks, sculpture etc.
Firebaskets and irons.
Prices: £10 to £10,000. Credit cards not
accepted
Note: Commissions undertaken

WORDSLEY ARTS & CRAFTS
19 Lawnswood Road
Wordsley
near **Stourbridge**
West Midlands DY8 5PG
Tel: 01384 278179
Fax: 01384 278179
Contact: Mrs. Ena Blood
Established: 1991
Open: 9am to 5.30pm Monday to
Saturday
Stock: Embroidery, découpage, art
products, art gallery, small gifts. Various

types of craft materials
Speciality: Cross stitch embroidery and limited edition prints
Prices: £1 to £40. Visa/Access/American Express accepted
Note: Occasional art exhibitions throughout the year (local artists). Framing centre in adjoining building by sister company 'Wordsley Framing'

PEDLARS PINE & CRAFT
Hollis House, The Square
Stow on the Wold
Gloucestershire GL54 1AF
Tel: 01451 832481
Fax: 01451 870501
Contact: R. Formby
Established: 1989
Open: 9.30am to 5.30pm Monday to Saturday. 11am to 5pm Sunday
Stock: Vast range of pine furniture & crafts
Speciality: Bespoke pine furniture
Prices: 20p to £1,000+. Visa/MasterCard/Amex/Switch/JCB accepted

TRAFFORDS
Digbeth Street
Stow on the Wold
Gloucestershire GL54 1BN
Tel: 01451 830424
Contact: Mr. and Mrs. De Trafford
Established: 1994
Open: Monday to Saturday 9am to 6pm. Sunday 10am to 5pm
Stock: High quality British contemporary crafts, domestic pottery, studio ceramics, jewellery, wood, metal, textiles, glass and baskets
Prices: From under £5 to £400. Major credit cards accepted
Note: Situated in pretty Cotswold town

CORIOLANUS CRAFTS & GIFTS
2 Centre Craft Yard, Henley Street
Stratford-upon-Avon
Warwickshire CV37 6QW
Tel: 01789 292901
Contact: Michelle and Keith Wilkinson
Established: 1994
Stock: Handmade cards, ornamental candles, handmade jewellery & tapestries
Speciality: Quality hand-crafted gifts and cards made in Great Britain
Prices: £1 to £200. Access/Visa/Mastercard/Delta/Switch/JCB accepted
Note: 50 metres from William Shakespeare's birthplace

MONTPELLIER GALLERY
8 Chapel Street
Stratford-upon-Avon
Warwickshire
CV37 3EP
Tel: 01789 261161
Contact: Linda Burridge
Established: 1992
Open: Monday to Saturday 9.30am to 5.30pm
Stock: High quality contemporary ceramics, glass and jewellery, paintings, printmaking and sculpture
Prices: £20 to £1500. Switch/Delta
Note: Exhibition programme throughout year

NICHOLAS WOOD FURNITURE MAKER
The Brickyard
Preston-on-Stour
Stratford-upon-Avon
Warwickshire
CV37 8BN
Tel: 01789 450883
Fax: 01789 450100
Contact: Patrick Hill and Graham Wood
Established: 1982
Open: Monday to Friday 9am to 5pm, Saturday 9am to 4pm or by appointment
Craftworkers on site: 10
Crafts: Traditional English furniture in English hardwoods. Restoration and commission work
Prices: £25 to £3,000. Credit cards not accepted
Note: Visitors can see raw timbers being machined and the whole fascinating process of crafting furniture

THE CRAFT GALLERY
33 Henley Street
Stratford-upon-Avon
Warwickshire
CV37 6QW
Tel: 01789 296996
Contact: Tooty Hyett and Carol Moss
Established: 1983
Open: Summer - 10am to 5pm daily. Winter - 10.30am to 4pm. Closed Mondays
Stock: High quality crafts both modern and traditional, hand made locally
Prices: 50p to £500. Visa/Mastercard/Eurocard/JCB accepted
Note: 30 yards from Shakespeare's birthplace

ARBORIAN
Unit 10, Piccadilly Mill
Lower Street
Stroud
Gloucestershire GL5 2HT
Tel: 01453 753287
Contact: Ian D.F. Sim
Established: 1986
Open: Please telephone in advance
Crafts: Small wooden items/furniture.
Candleholders and angels etc.
Prices: £2 to £300+. No credit cards

THE WAREHOUSE
50 High Street
Stroud
Gloucestershire GL5 1AN
Tel: 01453 752439
Contact: Jamie Lloyd and Nichola Harris
Established: 1990
Open: 9.30am to 5.30pm Monday to
Saturday
Stock: Ceramics, carvings, rugs, hats,
bags, clothing, exotic furniture and
jewellery etc.
Prices: £1 to £300. Access/Visa/
Mastercard/Eurocard accepted
Note: Situated in small beautiful
Cotswold town our shop prides itself in
the unusual stock

YEW TREE GALLERY
Steanbridge Lane, Slad
near **Stroud**
Gloucestershire GL6 7QE
Tel: 01452 813601
Contact: Gill Wyatt Smith
Established: 1971
Open: During exhibitions Tuesday to
Saturday 10.30am to 5.30pm, Sunday
2pm to 5pm. Please 'phone at other times
Stock: High quality domestic pottery,
studio ceramics, wood, glass, metal,
jewellery, textiles, furniture as well as
paintings and original prints
Prices: Under £25 to over £100. Major
credit cards accepted
Note: Five rooms of Cotswold house in
beautiful setting. Variety of contemporary
British craft and paintings. Four exhibitions held each year

MIDDLETON HALL CRAFT CENTRE
Middleton Hall, Middleton
Drayton Manor Park
near **Tamworth**
Staffordshire B78 2AE

Tel: 01827 261439
Contact: Margaret Hill
Established: 1990
Open: 11am to 5pm Wednesday to
Sunday
Craftworkers on site: 10
Crafts/stock: Artist, architectural and
decorative plasterer, ceramic sculptor,
silk painted textiles, upholsterer and
antiques, picture framer and prints,
wooden garden furniture, children's
furniture and toys, silversmith
Prices: £2 to £1,500. Visa/Mastercard
accepted
Note: Workshops/tutorials and demonstrations. Hall open Sundays April to
October 2pm to 5pm. Range of events.
Courtyard coffee shop, home cooked
meals and cakes, historic buildings, lake,
walks, country setting. Limited radar lock
facilities for disabled

ECCENTRIC
Park House, off Park Street
Madeley
Telford
Shropshire TF7 5HE
Tel: 01952 583345
Fax: 01952 583345
Contact: John Linder
Established: 1987
Open: 9am to 1pm and 2pm to 5.30pm.
Wednesday 9am to 1pm. Saturday 9am
to 5.30pm. Other times by arrangement
Stock: Pottery, jewellery, brassware, dried
flowers
Speciality: Meccano sets and parts new
and used. Scout and Guide wear and
equipment
Prices: 50p to £30. Barclaycard/Access
accepted
Note: One of the largest stocks of
Meccano parts and sets in UK.
Secondhand No. 10 sets a speciality at
around £1,000

NUMBER NINE
9 Tontine Hill
Ironbridge
Telford
Shropshire TF8 7AL
Tel: 01952 432344
Contact: Jayne Mountford Jones
Established: 1994
Open: 10am to 5pm daily Easter to
October 31st. Closed Mondays during
winter

Stock: Furniture, soft furnishing, dhurries, wide selection of candles. Bronze, silver celtic jewellery, framed prints and giftware
Speciality: Hand painted decorative furniture
Prices: 95p to £250. All major credit cards accepted
Note: Situated opposite the world famous Ironbridge

THE OLD STABLES GALLERY
The High Street
Ironbridge
Telford
Shropshire
TF8 7AD
Tel: 01952 433158
Contact: Elaine Shirley
Established: 1992
Open: 11am to 5pm daily. Closed Wednesday and Thursday
Stock: Contemporary ceramics, metalwork, art, glass and jewellery all hand crafted by top makers from throughout the British Isles
Speciality: Unusual candlesticks
Prices: £5 to £500. Visa/Mastercard
Note: Periodic themed exhibitions held. Short walk from the famous Iron Bridge

THE CRAFT SHOPPE
1 & 2 Teme Court Arcade, Teme Street
Tenbury Wells
Worcestershire
WR15 8AA
Tel: 01584 811757
Contact: Mrs. Susan Lane
Established: 1992
Open: 10am to 4.30pm Monday to Saturday. Some Sundays during summer. Closed Mondays January to Easter
Stock: Canal ware, salt dough, ceramics, encaustic art, wood turning, candles, stained glass, unusual flower decorations, items captured in copper. All items are original and exclusive designs
Prices: £1 to £100. Visa/Mastercard accepted

CHICO'S CRAFTS
21A Market Place
Tetbury
Gloucestershire
GL8 8DD
Tel: 01666 504493
Fax: 01249 657282

Contact: John Warner
Established: 1990
Open: Friday and Saturdays
Stock: Hand crafted original woodcarvings of wildfowl and other decoy. Rechargeable bamboo (paraffin) garden torches
Prices: £2.50 to £30. No credit cards

HOOKSHOUSE POTTERY
Westonbirt
Tetbury
Gloucestershire GL8 8TZ
Tel: 01666 880297
Contact: Christopher White
Established: 1975
Open: 10am to 6pm incl. Saturday and Sunday
Craftworkers on site: 1
Crafts: High-fired stoneware. One-off pieces, domestic ware and garden pots
Prices: £1.85 to £150. Credit cards not accepted but own credit scheme available
Note: Twice yearly mixed craft/art exhibitions and courses held

CONDERTON POTTERY
The Old Forge
Conderton
near **Tewkesbury**
Gloucestershire GL20 7PP
Tel: 01386 725387
Contact: Toff Milway
Open: 9am to 5pm Monday to Saturday and other times by appointment
Prices: Affordable workshop prices
Crafts: Stoneware pottery much of it salt glazed. African influence can be seen in finely textured surfaces and roulette decorations of many of the pots

THE BLAKESLEY GALLERY
Barton House
High Street, Blakesley
Towcester
Northamptonshire
NN12 8RE
Tel: 01327 860282
Fax: 01327 860282
Contact: George and Romayne Wisner
Established: 1979
Open: 10am to 5pm Wednesday to Saturday. Closed 25th December to mid-January
Stock: Ceramics, glass, textiles, jewellery, sculpture - both indoor and outdoor.

Wood, metalwork, watercolours
Speciality: Watercolours. Four mixed
exhibitions held each year
Prices: £5 to £1,000. Visa/Access/
Mastercard
Note: Crafts Council selected. Gallery
housed in 400 year-old barn in conserva-
tion area. Rural location. Free parking.
Pub next door

J. F. SPENCE & SON
The New Forge
Station Road
Uppingham
Rutland LE15 9TX
Tel: 01572 822758
Fax: 01572 821348
Contact: Mr. Spence
Established: 1896
Open: 8am to 12.30pm, 1.30pm to 5pm
Monday to Friday. 9am to 12.30pm
Saturday
Craftworkers on site: 4
Crafts: Dog grates, fire baskets, gates,
railings, garden ironwork etc.
Note: Teashops, antique shops in
Uppingham, a small old market town full
of character.

MAGPIE GALLERY
2 High Street West
Uppingham
Rutland LE15 9QD
Tel: 01572 822212
Contact: Roger Porter
Established: 1994
Open: Monday to Saturday 9am to
5.30pm
Stock: High quality domestic pottery,
studio ceramics, jewellery, wood, metal,
textiles, glass, baskets, paper, leather,
cards as well as original paintings/prints
Prices: From £5 to £1,000. Major credit
cards accepted
Note: Framing service available.
Changing displays of contemporary crafts
by new and established makers. Centrally
situated in picturesque market town

WALSALL LEATHER MUSEUM
Wisemore
Walsall
West Midlands
WS2 8EQ
Tel: 01922 721153
Contact: M. Glasson
Established: 1988

Open: Tuesday to Saturday 10am to 5pm.
(November to March closed at 4pm)
Craftworkers on site: 4
Crafts/stock: Small leather goods (book
marks, key fobs, stud boxes etc.). Large
shop selling range of Walsall-made bags,
belts, purses etc.
Prices: 50p to £95. Credit cards not
accepted
Note: Daily demonstrations, historical
display. Cafe, gardens, shop. Disabled
access

HATTON COUNTRY WORLD
Dark Lane, Hatton
Warwick
Warwickshire CV35 8XA
Tel: 01926 843411
Fax: 01926 842023
Contact: Carol Blower
Established: 1982
Open: 9am to 5.30pm daily except
Christmas day
Stock: Soft furnishings, wooden toys,
puzzles and gifts, stencils and accesso-
ries, designer fabrics, dried silk and
pressed flowers, stoneware pottery,
woodturning, ceramics, candles and
soaps, traditional craft metalwork and
copper jewellery, printers, stationers and
goldblocking specialists, music boxes and
porcelain dolls, wrought iron, silverware,
hand engraved glass, crystal and mirrors,
leather footwear, stained & painted glass
Note: Disabled access

CRAFT & HOBBY WORLD
17e Silver Street
Wellingborough
Northamptonshire NN8 5FE
Tel: 01933 271977
Fax: 01933 679367
Contact: Gill Weightman
Established: 1990
Open: 9am to 5.30pm Monday to
Saturday. Thursday 9am to 1pm
Stock: Paintings and photos by local
artists, casting crafts, dolls houses, papier
mâché, jewellery, glass, wooden items,
candles. Extensive range of craft and
hobby kits and materials
Prices: £2 to £100. Visa/Mastercard

JANE LANGAN – DESIGNER JEWELLER
The Castle Arts Centre
Castle Way
Wellingborough

Northamptonshire NN8 1XA
Tel: 01933 229022
Fax: 01933 229888
Contact: Jane Langan
Established: 1995
Open: 10am to 6pm daily
Craftworkers on site: 1
Crafts: Unusual contemporary silver designer jewellery. Precious jewellery to commission
Prices: £10 to £600. Visa/Mastercard/Switch/Delta
Note: Jewellery workshops and evening classes held including dance, drama, printing workshops. Art gallery, theatre, bar, outside eating area. Disabled access to all areas except silversmith workshop

THE ART IN ACTION GALLERY
Waterperry Gardens
Waterperry
near **Wheatley**
Oxfordshire
OX33 1JZ
Tel: 01844 338085
Fax: 01844 339883
Contact: Pat Clayton
Established: 1995
Open: March-October: 10am to 5.30pm Monday to Friday and 10am to 6pm Saturday and Sunday. November-February: 10am to 5pm
Stock: Ceramics, glass, etchings, jewellery, paintings, textiles, wood engravings, woodturning. Every craft hand made to the highest standard
Prices: £5 to £1500. Mastercard/Visa/Switch accepted
Note: Special exhibitions with craft demonstrations. Art in Action 4 day event held in July each year with over 250 demonstrating. Gardens of 83 acres with shop/nursery. Teashop - everything on menu homemade

SHARON WEBB
Gladstone Villas
56 North Street
Whitwick
Leicestershire LE67 5HA
Tel: 01530 811708
Contact: Sharon Webb
Open: By appointment only
Craftworkers on site: 1
Crafts: Textiles/interior designs including hand painted recycled glassware ornamental and functional

Prices: £5 to £30. Credit cards not accepted
Note: Various 'interior design' items are produced including curtains and co-ordinates by commission. Workshops - inc. certificated (LOCN) accredited interior/textile courses by arrangement

WINCHCOMBE POTTERY
Broadway Road
Winchcombe
near Cheltenham
Gloucestershire GL54 5NU
Tel: 01242 602462
Contact: Mike Finch
Established: 1926
Open: 8am to 5pm Monday to Friday, 10am to 4pm Saturday, Summer 12 noon to 4pm Sundays
Craftworkers on site: 5
Crafts: Hand made wood fired stoneware domestic pottery. Sculptor and furniture maker
Prices: £2 to £100. Visa/Mastercard/Access accepted
Note: Shop and workshops open to public (no admission charge)

THE HOP POCKET CRAFT SHOP
The Hop Pocket
New House
Bishop's Frome
Worcester
Worcestershire
WR6 5BT
Tel: 01531 640323
Fax: 01531 640684
Contact: Mrs. Janet Pudge and John Pudge
Established: 1988
Open: March to Xmas Eve Tuesday to Saturday 10.30am to 5.30pm. Closed Mondays (except Bank Holidays and December) Sunday 2.30pm to 5pm. January and February Friday and Saturday 10.30am to 5pm, Sunday 2.30pm to 5pm
Stock: Clothes, paintings, prints, pottery, ceramics, wood turning and furniture, jewellery, hats, waistcoats, plants, terracotta pots
Speciality: Hop garland/vines and hop bunches and pillows
Prices: 20p to £250. Access/Visa accepted
Note: Tearoom, garden with topiary (yew bowers). Tours of hop kilns and hop yard by appointment. Easy parking

ENGLISH COUNTRY POTTERY
Station Road
Wickwar
Wotton-under-Edge
Gloucestershire
GL12 8NB
Tel: 01454 299100
Fax: 01454 294053
Contact: Julia Scull, Pam Hyde & John
Collett
Established: 1977
Open: Showroom 8.30am to 4.30pm
Monday to Friday. Workshop last 2
Sundays November and first 2 Sundays
December
Craftworkers on site: 30
Crafts: Hand painted pottery
Prices: £1.95 to £45. Visa/Eurocard/
Mastercard accepted
Note: Free tours of workshop and chance
to paint your own pottery last 2 Sundays
in November and first 2 in December.
Shared craft shop in Bath - 16 Cheap
Street

HAREHOPE FORGE POTTERY
Harehope Farm
Eglingham
Alnwick
Northumberland NE66 2DW
Tel: 01668 217347
Fax: 01665 510624
E-mail: Richard@tag.co.uk
Contact: Richard Charters
Established: 1990
Open: Please telephone in advance
Craftworkers on site: 1
Crafts: Handmade woodfired terracotta
garden pots and glazed kitchen pots
Prices: £5.50 to £500. Credit cards not
accepted
Note: Demonstrations given to pre-
arranged parties. Wholesale trade is
welcome

WOODHORN FURNITURE
Woodhorn Colliery Museum
QEII Country Park
Ashington
Northumberland
NE63 9YF
Tel: 01670 856592
Contact: R. Blake
Established: 1993
Open: 8am to 4pm Wednesday to
Monday. Closed Tuesday
Craftworkers on site: 1
Crafts: Handmade furniture to order
Note: Matmaking workshop, flute maker,
guitar maker, blacksmith, cafe also on site

MIDDLETON CRAFTS
1c Chapel Row
Middleton-in-Teesdale
Barnard Castle
County Durham DL12 0SN
Contact: Janet Rawle
Established: 1986
Open: April to October 10am to 5pm
daily. November to March Friday,
Saturday and Sunday only 11am to 4pm
Stock: Dry stone wall models, landscape
photographs, needlework, pressed wild
flower cards, painting china and glass,
soft toys, stone clocks, dried and silk
flower arrangements, animal models,
jewellery, wood turning, papier-mâché
and more
Note: Informal co-operative. Community
based project for local crafts-people

PRIORS CONTEMPORARY FINE ARTS AND CRAFTS
First Floor, 7 The Bank
(above Oldfields Restaurant)
Barnard Castle
County Durham DL12 8PH
Tel: 01833 638141
Contact: Mark Prior
Established: 1985
Open: Monday to Friday 10am to 5pm.
Saturday 10am to 5.30pm. Sunday 12
noon to 5.30pm
Stock: Paintings, prints, ceramics,
jewellery, wood, glass, landscape
photography, books and cards. Ecological
framing service and moulding supply
Note: Located in unusual Dales market
town with castle and museum

STUDIO TWO
Norselands Gallery
The Old School, Warenford
near **Belford**
Northumberland NE70 7HY
Tel: 01668 213465
Contact: Veronica Rawlinson
Established: 1979
Open: 9am to 4pm daily (Gallery 9am to
6pm)
Craftworkers on site: 2
Crafts/stock: Hand sculpted ceramic
figurines. High quality paintings, prints
and ceramics by mainly British makers
Prices: £1 to £200. Access/Visa/
Mastercard/JCB accepted
Note: Tea, coffee and homemade cakes,
children's play area, aviary and bonsai
house all available but on a very small
and attractive scale

TOWER HOUSE POTTERY
Tower Road
Tweedmouth
Berwick-upon-Tweed
Northumberland TD15 2BD
Tel: 01289 307314
Contact: P. Thomas
Established: 1976
Open: 10am to 5pm Monday to Friday
Craftworkers on site: 2-3
Crafts: Pottery
Prices: £1 to £1,500. Credit cards not
accepted
Note: Visitors may see workshop.
Children welcome

WINDMILL HOLE STUDIO
9 Railway Street
Berwick-upon-Tweed
Northumberland TD15 1NF
Tel: 01289 307135
Contact: Inger Lawrance
Established: 1989
Open: Tuesday and Friday 11am to 4pm
or by appointment
Craftworkers on site: 1 or 2
Crafts: Etchings, woodcuts, artists books,
ceramics, handcrafted papier-mâché
Prices: £15 to £500. Credit cards not
accepted
Note: Demonstrations by appointment of
printing, etching and/or woodblock
printing. Member of Northern Arts
Purchase Plan

BELT UP
Lelthamhill
Cornhill-on-Tweed
Northumberland
TD12 4TP
Tel: 01890 820227
Contact: Dave Downie
Established: 1988
Open: 10am to 5pm but please telephone
in advance
Craftworkers on site: 1
Crafts: Leather belts, handbags,
handsewn briefcases, buckles and other
items
Prices: £3.50 to £22. Credit cards not
accepted
Note: Belts fitted on the spot to customers'
needs

CRAFTS NINEHUNDRED
The Courtyard
Saddlers Yard, Saddler Street
Durham
DH1 3NP
Tel: 0191 386 3050
Contact: Angela Colbridge
Established: 1993
Open: 9.30am to 5pm Monday to
Saturday
Stock: Northumbrian crafts, wood, glass,
ceramics, pottery, textiles, toys, jams,
preserves, Durham mustard, furniture,
cards, jewellery
Prices: 50p to £250. All major credit
cards accepted
Note: Full programme of craft demonstra-
tions throughout the summer. Cafe in
courtyard

PORTCULLIS CRAFTS GALLERY
7 The Arcade
Metrocentre
Gateshead
Tyne & Wear
NE11 9YL
Tel: 0191 460 6345
Fax: 0191 460 4285
Contact: Wendy Turnbull
Established: 1989
Open: Monday to Wednesday and Friday
10am to 8pm, Thursday to 9pm. Saturday
9am to 7pm and Sunday 11am to 5pm.
Open Bank Holidays
Stock: High quality contemporary crafts
domestic pottery, studio ceramics, wood,
glass, metal, jewellery, toys, textiles and
knitting
Speciality: Ceramics and glass jewellery
Prices: From under £15 to over £1500
Most major credit cards accepted (Not
AmEx or Diners)
Note: Situated within the famous and
huge Metrocentre

THE LACE KITCHEN/QUILTS
25/27 The Boulevard
Antique Village
Metrocentre
Gateshead
Tyne & Wear
NE11 9YP
Tel: 0191 460 6264
Contact: Shirley Collins
Established: 1986
Open: 10am to 8pm Monday to Friday.
Thursday 10am to 9pm. Saturday 9am to
7pm. Sunday 11am to 5pm
Stock: Linen and lace, tablecloths,
bedspreads, patchwork quilts,
throwovers, porcelain dolls, lace gifts
Speciality: Patchwork quilts and lace
Prices: £1 to £300. Most credit cards
accepted
Note: Situated in Antique Village of
Metro Centre, Gateshead

PHOENIX HOT GLASS STUDIO
10A Penshaw Way, off Blackthorn Way
Sedgletch Ind. Est., Fencehouses
Houghton le Spring
Tyne & Wear
DH4 6JN
Tel: 0191 385 7204
Fax: 0191 385 7204
Contact: Mrs. Anne Tye
Established: 1989

Open: 10am to 1pm, 2pm to 5pm
Monday to Friday
Craftworkers on site: 5
Crafts: Mouthblown, handfinished
glassware
Prices: £8 to £120. Visa/Access/Connect
accepted
Note: Visitors are welcome to watch
glassmaking process. Good parking and
access for disabled. Groups must book in
advance

CLEVELAND CRAFTS CENTRE
57 Gilkes Street
Middlesbrough
TS1 5EL
Tel: 01642 262376
Fax: 01642 226351
Contact: Julia Palmer
Established: 1984
Open: Tuesday to Saturday 10am to 5pm
Stock: All crafts.
Speciality: Ceramics and contemporary
jewellery
Prices: £5 to £500
Note: Creative activities and courses
held. High quality temporary exhibition
programme. Contemporary non-precious
jewellery collection. Studio ceramics
collection

D.G. BURLEIGH
Rothbury Road
Longframlington
Morpeth
Northumberland
NE65 8HU
Tel: 01665 570635
Contact: D.G. Burleigh and O.F. Burleigh
Established: 1972
Open: 8am to 12 noon, 1.30pm to 4pm
Monday to Friday or by appointment.
Closed 25th December to 1st January
Craftworkers on site: 2
Crafts: Northumbrian smallpipes (musical
instrument). Music books, tapes, CDs
Prices: From about £430 to £850. Credit
cards not accepted
Note: Demonstrations held. Display of
photographs. Tea and coffee available

IONA ART GLASS
Woodlands
Warkworth
Morpeth
Northumberland
NE65 0SY

Tel: 01665 711533
Contact: C. Chesney
Established: 1986
Open: Most days but by appointment
only; please 'phone first
Craftworkers on site: 5
Crafts: Stained glass
Prices: £35 to £30,000 (commissions)
Note: These products are also available
from Northumbrian Makers at Blackfriars,
Friars Green, Stowell Street, Newcastle
upon Tyne – see over page for full details

NORTHUMBRIA CRAFT CENTRE
Morpeth Chantry
Bridge Steet
Morpeth
Northumberland
NE61 1PJ
Tel: 01670 511217
Fax: 01670 511326
Contact: Sue Jordan
Established: 1987
Open: 9.30am to 5.30pm Monday to
Saturday
Crafts/stock: Pottery, paintings, patch-
work, jewellery, woodturning, metal-
work, stained glass, embroidery, dry stone
models, glass, coal sculpture, clothes,
knitwear, hats, candles, honey,
souvenirs
Prices: £1 to £150. Visa/Switch/Delta
accepted
Note: Demonstrations held, artist in
residence, bagpipe museum, concerts,
tourist information centre. Disabled
access

ARABESQUE
205 Westgate Road
Newcastle upon Tyne
Tyne & Wear
NE4 6AB
Tel: 0191 222 1196
Contact: Margaret and Ernie
Established: 1990
Open: 10.30am to 5.30pm Monday to
Saturday
Stock: Craft goods from Africa, Indonesia,
India, Morocco, South America. Crystals,
oils, incense, pottery and silver
Speciality: Wood carvings, fantasy figures
Prices: £1.50 to £200. Credit cards
accepted
Note: Unusual one-off pieces made and
sold, ranging from jewellery to
sculptures

BONA DEA CANDLES
Unit 6
25 Low Friar Street
Newcastle upon Tyne
Tyne & Wear
NE1 5UE
Tel: 0191 233 0889
Fax: 0191 261 5746
Contact: Steven Dendle
Established: 1992
Open: 9.30am to 6pm Monday to Friday.
10.30am to 5pm Saturday
Craftworkers on site: 2
Crafts: Candles, paraffin wax, bees wax.
Cast, scented candles. Candle holders, oil
lamps, oil burners
Prices: £2.75 to £50. Credit cards not
accepted
Note: Candles made to order. All
handmade without animal products -
wholesale and retail available. Visitors
welcome during above opening hours,
please ring first

CLAYTON GALLERY
14 Clayton Road
Jesmond
Newcastle upon Tyne
Tyne & Wear
NE2 4RP
Tel: 0191 281 2560
Fax: 0191 281 6734
Contact: Margaret Walker
Established: 1992
Open: Monday to Saturday 10am to 5pm.
Thursday 10am to 6pm
Stock: Ceramics, glass, jewellery, textiles
and wood. Some design
Prices: £10 to £800
Note: Fine art contemporary exhibition -
changes every month with previews

NORTHUMBRIAN MAKERS AT
BLACKFRIARS
Friars Green
Stowell Street
Newcastle upon Tyne
Tyne & Wear
NE1 4XN
Tel: 0191 261 4307
Contact: Bryan Stanley
Established: 1995
Open: Monday to Saturday 10am to 5pm
Stock: Stained glass, jewellery, embroi-
dery, decorative ironwork, pottery, wood,
terracotta gardenware, leather,
handwoven jackets, scarves, handmade

clothing, handknits, painting
Prices: £2 to £300. Visa accepted
Note: Part of ancient friary complex of
Blackfriars. Restaurant adjacent. Disabled
access

OUSEBURN WAREHOUSE
WORKSHOPS
36 Lime Street
Byker
Newcastle upon Tyne
Tyne & Wear NE1 2PN
Tel: 0191 261 5993
Contact: John Fraser/David Wild
Established: 1992
Open: Most days 10am to 5pm but please
telephone to check
Craftworkers on site: 5
Crafts: Pottery, handthrown decorative
and functional stoneware including
bowls, vases and platters sold in shop.
Also on site glassworkers, furniture
makers and rag rug makers
Prices: £5 to £150. Credit cards not
accepted
Note: Visitors can ask to see around
workshops. Free parking. City Farm and
The Ship Inn are nearby

THE STAINED GLASS CENTRE
78A Forsyth Road
Jesmond
Newcastle upon Tyne
Tyne & Wear NE2 3EU
Tel: 0191 281 0945
Fax: 0191 281 4135
Contact: Greg Jones
Established: 1963
Open: Monday to Friday 10am to 5pm.
Saturday 9am to 12 noon. Closed all day
Wednesday
Craftworkers on site: 3
Crafts: Stained glass panels, lampshades,
anything glass!
Prices: £40 to £1,000. Access/Visa
accepted
Note: Evening classes arranged. The
unusual and difficult jobs welcomed.
1994 Aim High Awards - national
winners

THE EGG CRAFT STUDIO
7 Hylton Terrace
Coach Lane
North Shields
Tyne & Wear NE29 0EE
Tel: 0191 258 3648

Fax: 0191 258 3648
Contact: Joan Cutts
Established: 1975
Open: Please telephone for details
Craftworkers on site: 1
Crafts: Decorated eggs. Real eggs decorated in Fabergé style. Orders taken for any occasions
Prices: £6 to £50. Credit cards not accepted
Note: Seminars, weekend workshops, classes held. Talks for guilds can be arranged within 50 mile radius

THE FIREPLACE
Unit 1, Berwick Hill
Ponteland
Northumberland
NE20 0JZ
Tel: 01661 860154
Contact: Russell Elrington
Established: 1987
Open: Monday to Thursday 9am to 4.30pm, Friday 9am to 4pm, Saturday 10am to 3.30pm
Craftworkers on site: 1
Crafts: Wooden Victorian and Georgian style mantlepieces
Note: Fireplaces are made to order in any type of wood or style.There is also showroom to view many surrounds, marble, cast insets, brassware and fires

WESTFIELD FARM POTTERY AND GALLERY
Thropton
Rothbury
Northumberland
NE65 7LB
Tel: 01669 640263
Fax: 01669 640263
Contact: Catherine Hardie
Established: 1976
Open: Easter to Christmas, Tuesday to Saturday 10am to 5pm. Other times by appointment
Craftworkers on site: 2
Crafts/stock: Wood-fired kitchen and domestic stoneware. Jewellery, glass, fabrics, baskets, cards, fine prints etc.
Prices: £5 to £500. Visa/Mastercard
Note: Superb rural setting. Disabled access

DURHAM DALES CRAFTS
Durham Dales Centre
Stanhope

County Durham
DL13 2FJ
Tel: 01388 527650
Contact: The Chairman
Established: 1992
Open: 10am to 5pm daily. Telephone for opening times in January and February
Stock: Jewellery, découpage, ceramics, woodturning, painted glass, pewter craft, painting, stonecraft, needlework, pressed flowers, local scene tablemats and coasters and more
Prices: £1 to £150 or by commission. Credit cards not accepted
Note: A different craftsperson from the group staffs the shop each day. There are adjoining tearooms, tourist information centre and the castle gardens. Disabled access throughout the site including toilet facilities

WEAVERS
Forge Cottage
Ireshopeburn-in-Weardale
Stanhope
County Durham
DL13 1ER
Tel: 01388 537346
Contact: Michael and Mary Crompton
Established: 1978
Open: Summer 10am to 4pm. Winter by appointment
Craftworkers on site: 2
Crafts: Woven tapestries, handwoven fabrics, hand decorated bone china
Prices: £1 to £1,000
Note: Visitors can see traditional skills of hand loom weaver and tapestry weaver. Parking and access for disabled. Groups must book in advance

SUNDERLAND GLASS TRAIL
Sunderland Tourist Information Centre
Crowtree Road
Sunderland
Tyne & Wear
SR1 3EL
Tel: 0191 565 0960
Fax: 0191 565 3352
Contact: Kay Callaghan
Established: 1992
Open: Monday to Friday
Crafts/stock: Glassware.
Prices: £5 to £150. No credit cards
Note: This is not an outlet as such but an organised tour by bus with on-board courier visiting several glass workshops in

the area. Glass-blowing demonstrations
are seen at the stops and purchases can
be made at factory prices. Lunch stop on
the Tour that also visits other attractions
including Glorious Glass Exhibition at
Sunderland Museum and Art Gallery and
St. Peter's Church. Please phone first to
establish availability of places and
departure times

DIAL GALLERY
5 Dial Place
Warkworth
Northumberland
NE65 0UR
Tel: 01665 710822
Contact: Janice Charlton
Established: 1980
Open: Tuesday to Sunday 11am to
5.30pm. Closed Monday
Prices: From about £35 to over £1000.
Major credit cards accepted
Stock: Very high quality range of
contemporary crafts. Domestic pottery,
studio ceramics, wood, glass, metal,
jewellery and textiles as well as original
paintings and prints. This gallery is among
those currently listed by the Crafts
Council as of high quality

MINERVA HANDICRAFTS
27 Ainsworth Street
Blackburn
Lancashire BB1 1AA
Tel: 01254 664168
Contact: Pauline Davidson
Established: 1989
Open: 9.30am to 5pm Monday to
Saturday. Closed Thursdays
Stock: Needlecrafts and associated
products. Small range of numerous other
craft. Large selection for découpage
Speciality: Picture framing - needlework
and 3D
Prices: Pence to pounds. No credit cards

LUNA CRAFTS
51 Dickson Road
Blackpool
Lancashire FY1 2AT
Tel: 01253 27901
Contact: Ahmed Abo-Elazm
Established: 1975
Open: Summer - 10.30am to 6pm
Monday to Saturday. Winter - 10.30am to
6pm Thursday, Friday and Saturday.
Advisable to telephone
Stock: Egyptian crafts - Museum
reproductions, papyrus, Egyptian
jewellery, amulets, cards and crystals
Speciality: Egyptian crafts
Prices: 50p to £250. Access/Visa/Diners
Club/American Express/Mastercard
accepted
Note: Egyptian hieroglyphics on papyrus
to order

LAST DROP GALLERY
Last Drop Village
Bromley Cross
Bolton
Lancashire BL7 9PZ
Tel: 01204 593528
Contact: Mrs. Hurst and Mrs. Gill
Established: 1982
Open: 11am to 5pm daily. Closed
Christmas Day
Stock: Pewterware
Prices: 99p upwards. Most major credit
cards accepted
Note: Village complex. Shops, bakery,
teashop, restaurant, Drop Inn pub (buffet
lunches). Many attractions. Conference
centre. Extensive parking facilities - traffic
free village

THE MILESTONE
Last Drop Village, Bromley Cross
Bolton
Lancashire BL7 9PZ
Tel: 01204 305818
Contact: Judith and Harry Holt
Established: 1982
Open: Monday to Friday 11.30am to
5pm. Saturday and Sunday 10am to 5pm
Stock: Gemstone jewellery, mineral
specimens, silver, gold and gem sets.
Jewellery by some of Britain's best
designers. Commissions accepted
Speciality: Vast range of stones, cut and
polished in our own workshop
Prices: £5 to £500 Visa, MasterCard/
Amex/Switch accepted
Note: Part of 'Last Drop Village' – see
previous entry

SLATE AGE (FENCE) LTD.
Fence Gate, Fence
near **Burnley**
Lancashire BB12 9EG
Tel: 01282 616952
Fax: 01282 619058
Contact: Mrs. Kathryn Rawlinson
Established: 1969
Open: 8am to 4.30pm Monday to Friday.
9am to 1pm Saturday
Craftworkers on site: 6
Crafts: Slate giftware: clocks, thermom-
eters, desk sets, book ends, novelties etc.
Pottery, basketware, lace, cards, notepaper
Prices: 50p to £40. Visa/MasterCard/
Access accepted
Note: Workshops open to public.

LONGNOR CRAFT CENTRE
The Market Hall
Market Square, Longnor
Buxton
Derbyshire SK17 0NT
Tel: 01298 83587
Contact: Carol Wheeldon
Established: 1991
Open: 10am to 5pm daily. Weekends
only first 10 weeks of year
Stock: Locally made hardwood furniture,
pottery, woodturnery, paintings,
patchwork, embroideries
Speciality: Hardwood and country furniture
Prices: £3 to £2,500. No credit cards
Note: Monthly changing exhibitions.
Coffee shop

ROOKES POTTERY
Mill Lane
Hartington
Buxton
Derbyshire SK17 0AN
Tel: 01298 84650
Fax: 01298 84806
Contact: David Rooke
Established: 1977
Open: 9am to 5pm Weekdays. 10am to
5pm Saturday. 11am to 5pm Sunday
Craftworkers on site: 3 full-time and 3
part-time
Crafts: Original range of terracotta garden
pots
Prices: 65p to £90. Access/Visa/Switch
accepted
Note: Seconds always available -
workshop open to visitors. Situated in the
limestone village of Hartington with duck
pond and several teashops. Good
disabled access

TOBILANE DESIGNS
Newton Holme Farm
Whittington
Carnforth
Lancashire LA6 2NZ
Tel: 01524 272662
Fax: 01524 272662
Contact: Elaine and Paul Commander
Established: 1985
Open: 10am to 5pm Monday to Saturday
Stock: Traditional wooden toys: farms,
castles, forts, dollshouses, rocking horses
and sheep. Also educational jigsaws, soft
dolls, collector teddies and old fashioned
wooden toys
Speciality: Restoration of old toys
including rocking horses
Prices: 75p to £1300
Note: Restoration and new work in
progress can often be seen. Limited
disabled access as no ramp to shop.

CHESHIRE WORKSHOPS LTD.
Burwardsley
Tattenhall
Chester
Cheshire CH3 9PF
Tel: 01829 770401
Fax: 01829 770440
Contact: Graeme Fell
Established: 1974
Open: 10am to 5pm seven days a week
Craftworkers on site: 16
Crafts: Carved candles, tapered candles,
floating candles. Glass sculpture, wood
turning, pyrography
Prices: 1p to £50. All credit cards
accepted
Note: Restaurant, food hall, animals.
Disabled access

CONSTELLATION
70 Lower Bridge Street
Chester
Cheshire CH1 1RU
Tel: 01244 316083
Fax: 01244 316083
Contact: Stella Abbotson
Established: 1989
Open: 10am to 6pm Monday to Friday.
9.30am to 5.30pm Saturday. 1pm to 5pm
Sunday
Stock: Beads and jewellery plus other
crafts
Prices: £1 to £30. Visa/Mastercard
accepted
Note: Bead jewellery classes held. Close
to city wall and river

THE CHESTER CANDLE SHOP
75 Bridge Street Row
Chester
Cheshire CH1 1NW
Tel: 01244 346011
Fax: 0151 652 5931
Contact: Mrs. Ghislaine Payne
Established: 1984
Open: 9.30am to 5.30pm weekdays and
Bank Holidays. 11am to 4pm Sundays
Craftworkers on site: 7
Crafts: Hand carved candles, candle-
rings, candlesticks and holders and other
gift items
Prices: 8p to £40. Credit cards accepted
Note: Demonstrations of candles being
made and skilled carvers in action

TOP FARM POTTERY
High Street, Farndon
Chester
Cheshire CH3 6PT
Tel: 01978 364812
Contact: Sue and Willie Carter
Established: 1981
Open: 9am to 5pm
Craftworkers on site: 1
Crafts: Wide range of functional
stoneware decorated with fish, chickens
and geese. Other designs also available
Prices: £7 to £250. Credit cards not
accepted

GILLIES STUDIO
8 Bolton Road
Adlington
near **Chorley**
Lancashire PR6 9NA
Tel: 01257 480637
Fax: 01257 482074
Contact: Mrs. G. Charlson
Established: 1975
Open: 1.30pm to 5pm daily. Closed
Wednesdays and Sundays
Craftworkers on site: 2
Crafts/Stock: Artist dolls - porcelain and
wax, ceramics. Dollmaking and ceramic
supplies
Note: Portraits in porcelain made from
photos. Antique restoration of wax and
porcelain dolls. Classes in dollmaking
and ceramics

HANDMADE CERAMIC PRODUCTS
Rose Cottage
150 Biddulph Road, Mossley
Congleton
Cheshire CW12 3LY
Tel: 01260 277237
Contact: Cynthia Pulman
Established: 1982
Open: Most days but telephone first
Craftworkers on site: 3
Crafts: Handmade and slipcast animals
and birds: pigs, horses, cats, owls etc.
Prices: under £1 to £30

MOUNT CRAFT CENTRE
The Esplanade
Fleetwood
Lancashire
Tel: 01253 884242
Contact: Mrs. M. Wolstenholme
Established: 1986
Open: April-May 12.30am to 5pm.
Summer 10.30am to 5pm daily. Shorter
hours in winter
Craftworkers on site: 35
Crafts/stock: Soft toys, watercolours,
cross stitch, crochet, quilting, jewellery,
dried flower arrangements, pressed flower
pictures, wood turning, candles, 3D
découpage, decorated eggs, honey
products, handpainted wooden goods,
pottery, household software, books
Prices: 10p to £250. Visa/Access/
MasterCard accepted
Note: Craft centre housed in Grade 2
listed building. Exhibition in summer by
Civic Society of Fleetwood

STUDIO ARTS GALLERY
6 Lower Church Street
Lancaster
Lancashire LA1 1NP
Tel: 01524 68014
Fax: 01524 844422
Contact: Iain Dodgson
Established: 1971
Open: 9am to 5.30pm Monday to
Saturday
Stock: Limited edition prints, originals,
ornaments, sculptures
Prices: £5 to £50,000. All major credit
cards accepted

BLUECOAT DISPLAY CENTRE
Bluecoat Chambers, School Lane
Liverpool
Merseyside L1 3BX
Tel: 0151 709 4014
Contact: Maureen Bampton
Established: 1959
Open: 10.30am to 5.30pm Monday to
Saturday
Stock: Ceramics, studio pottery, glass,
precious jewellery, costume jewellery,
textiles, metalwork, paperwork, wood
Speciality: Ceramics
Prices: £5 to £800. Access/Visa/
Mastercard/American Express/Switch/
Delta accepted
Note: Unusual and varied exhibition
programme. Six exhibitions held every
year featuring different media. Non-profit
making organisation on Crafts Council's
list of selected galleries. Located in arts
centre in beautiful Queen Anne building
beside lovely courtyard garden. Cafe, bar
and other shops

GUANGMING COMPANY
22 Berry Street
Liverpool
Merseyside L1 4JF
Tel: 0151 708 9235
Fax: 0151 709 0967
Contact: John Sung
Established: 1978
Open: Monday to Saturday 9.30am to
5.30pm. Sunday 12 noon to 5.30pm
Stock: Chinese arts and crafts including
porcelain figures, prints, jewellery,
pictures, handicrafts. Wide range of
Chinese brush painting materials
Prices: £2 to £100. No credit cards
Note: Located in heart of Liverpool's
Chinatown. Mail order service available

WALKER ART GALLERY SHOP
William Brown Street
Liverpool
Merseyside
L3 8EL
Tel: 0151 207 0001
Open: 10am to 5pm Monday to Saturday
and 12pm to 5pm on Sunday
Stock: High quality designer crafts with
an emphasis on those made locally.
Ceramics, jewellery, silverwork, mirrors,
glass etc. Mail order also possible
Prices: From under £5 to over £100.
Credit Cards accepted
Note: This craft shop is located within the
famous Walker Art Gallery

BUBBLE CRAFT SHOP
The Square, St. Annes
Lytham St. Annes
Lancashire
Contact: Mrs. S. Higham
Established: 1995
Open: 10am to 5pm most days except
Mondays
Stock: Flower arrangements and crafts
Prices: £1 to £150. Credit cards not
accepted
Note: Craft demonstrations held. Situated
in an attractive seaside resort

DRAGONFLY DECORATION AND DESIGN
Church Farm
Marton
near **Macclesfield**
Cheshire
Tel: 01260 277561
Contact: Peter and Rowan Webb
Established: 1992
Open: Every weekend and Tuesday,
Wednesday and Thursday afternoons
Craftworkers on site: 2
Crafts: Pictures, decorated furniture,
antiques, pottery, glass, etc.
Prices: Various
Note: Courses available in decorative
paint finishes - one day course £49.
Teashop, gardens, farm animals etc.

CRAFTS & THINGS
126 Mauldeth Road
Fallowfield
Manchester M14 6SQ
Tel: 0161 448 0080
Contact: D. Wilkinson
Established: 1991

Open: 10am to 5pm. Closed Wednesdays
Stock: Ethnic carvings, baskets, dried
flowers, handmade cards, giftware, hats,
rugs, pictures, candles, oils and incense
Speciality: Ethnic items
Prices: £1 to £90. Most credit cards
accepted

MANCHESTER CRAFT CENTRE
17 Oak Street
Smithfield
Manchester M4 5JD
Tel: 0161 832 4274
Fax: 0161 832 3416
Contact: Janine Hague
Established: 1982
Open: Monday to Saturday 10am to
5.30pm. Closed Bank Holidays
Stock: Ceramics, glass, jewellery, wood,
textiles, sculpture, handpainted silk
products and designer clothing including
hand made ties, greetings cards
Prices: £8 to £250+. Visa accepted
Note: Exhibitions held year round,
demonstrations, cafe. Situated in restored
Victorian market hall with original glass
roof. Disabled access ground floor & WC

NAMASTE VILLAGE
1B School Lane, Didsbury
Manchester M20 0RD
Tel: 0161 445 8230
Contact: Sue Kagan
Established: 1993
Open: 10am to 5.30pm all year
Stock: Exotic gift items from around the
world. Candles and burners, carved wood
and handpainted ceramics, windchimes
and interesting jewellery. Cards and
paintings
Speciality: International selection of
silver jewellery and Mexican furniture
made from reclaimed timbers
Prices: £1 to £200. Visa/Access/Amex
accepted (Switch not accepted)
Note: Excellent village atmosphere with
other craft shops, cafes and tearooms
locally. Access for wheelchairs and prams

ROYAL EXCHANGE THEATRE CRAFT CENTRE
Royal Exchange Theatre
St Ann's Square
Manchester
M2 7DH
Tel: 0161 833 9333
Fax: 0161 832 0881

Open: 10am to start of first performance. Closed Sunday
Stock: High quality domestic pottery, studio ceramics, glass, wood, jewellery, textiles, toys
Note: Unusual glass structure in the foyer of theatre promotes new and established makers through regular stock and exhibitions

THE FIRS POTTERY
Sheppenhall Lane, Aston
Nantwich
Cheshire CW5 8DE
Tel: 01270 780345
Contact: Joy Wild
Established: 1983
Open: 9am to 6pm daily incl. Saturdays and Sundays. Closed some Thursdays
Craftworkers on site: 2
Crafts: Stoneware pottery, functional and decorative ware
Prices: £2 to £70
Note: Pottery classes held. "Have a Making Day" - 10am to 4.30pm incl. lunch, tea, coffee etc. Booking essential. Children's workshops in holidays (half day). Gardens open in summer "See Pottery in a Garden Setting"

BARTINGTON FORGE
Warrington Road, Bartington
near **Northwich**
Cheshire CW8 4QU
Tel: 01606 851553
Fax: 01606 851553
Contact: David Wilson
Established: 1982
Open: 8am to 5pm Monday to Friday. 9.30am to 12.30pm Saturday
Craftworkers on site: 3
Crafts: High quality wrought ironwork - gates, railings, fire furniture, weather vanes. Brass and copperware
Prices: £1 upwards
Note: Member of National Association of Master Farriers, Blacksmiths and Agricultural Engineers

ARTISAN
3 King Street
Delph, Saddleworth
near **Oldham**
Lancashire OL3 5OL
Tel: 01457 874506
Contact: Anne Hamlett
Established: 1980

Craftworkers on site: 1+
Open: 1pm to 5.30pm Wednesday, Thursday, Friday. 10.30am to 5.30pm Saturday. Sunday and Bank Holidays 2pm to 5pm
Craft/Stock: Pottery, fantasy ceramics, castles, dragons, wizards etc. Oil burners and candle holders. Domestic pottery, glass, prints and paintings
Prices: 95p to £350. Access/Visa/Mastercard accepted
Note: Pottery making demonstrations but no set times. Member of Northern Potters Association

SADDLEWORTH CRAFTS (CO-OP) LTD.
12 King Street
Delph
Oldham
Lancashire OL3 5DQ
Tel: 01457 874705
Contact: Mrs. M. Eastham
Established: 1979
Open: All year except Christmas and New Year
Stock: Aromatherapy, art, crochet and tatting, dried flowers, embroidery. Gift baskets, engraved glass and tiffany stained glass. Greetings cards. Jams and preserves. Knitting, patchwork, picture framing, pottery, sewing, woodcrafts and toys. Wrought iron work
Prices: 75p to £125. Credit cards not accepted
Note: Art gallery and teashop on premises

THE ALEXANDRA CRAFT CENTRE
High Street
Uppermill
Oldham
Lancashire OL3 6HT
Tel: 01457 876675
Fax: 01457 876675
Contact: Mrs. Chorlton
Established: 1980
Open: 11am to 4.30pm Wednesday to Sunday inclusive
Stock: Knitwear, toy making, dressmaking, furniture making, handmade curtains, cushions, china, stained glass, picture framing, handmade beds, kitchen units, Christmas shop
Prices: £5 to £5,000. Credit cards not accepted
Note: Centre has two cafes. Situated in lovely walking country, the village has museum and working canal boat

THE ARTS & CRAFT CENTRE
15 Bow Street
Oldham
Lancashire
OL1 1SJ
Tel: 0161 628 7868
Contact: Mrs. Kate Hancock
Established: 1994
Open: Monday to Friday 10am to
5.30pm. Saturday 9.30am to 5pm
Crafts/stock: Jewellery, woodwork,
ceramics, needlework, toys, glassware,
salt dough, glass etchings, calligraphy,
dried flowers, 3D découpage, quilling,
greetings cards, knitwear, dolls houses,
porcelain dolls. Paintings and drawings
(all local artists and crafts people)
Prices: £1.50 to £700. Credit cards not
accepted
Note: Workshops in various crafts and
art classes held. Craft materials shop and
coffee shop

'THE OLD POST OFFICE' CRAFT CENTRE
57 School Lane, Haskayne
near **Ormskirk**
Lancashire L39 7JE
Tel: 01704 841066
Contact: Mrs. Kathy Cunliffe
Established: 1987
Open: Thursday, Friday and Saturday
9.30am to 4.30pm
Stock: Wooden toys, woodturnery
including miniatures. Exclusive doll
houses, furniture and accessories.
Handpainted silks, canal art, jewellery
(including silver, enamel, wooden),
handpainted china, cards, cushions, cot
covers
Prices: 15p to £1,000. Access/Visa/
Mastercard/Eurocard accepted
Note: Hairdressers and sunbed on
premises. Adjacent premises: pine
furniture, kitchens and bedrooms

ARTIZANA & ARTIZANA FURNITURE
The Village
Prestbury
near Macclesfield
Cheshire
SK10 4DG
Tel: 01625 827582
Fax: 01625 827582
Contact: Mrs Jemila Topalian Ghazoul
Established: 1984
Stock: High quality domestic pottery,

studio ceramics, glass, wood, jewellery,
textiles, toys, but particularly furniture
Prices: From under £20 for glass to over
£5000 for furniture. All major credit cards
accepted plus Switch
Note: Contemporary British crafts with a
separate section dedicated to latest work
of leading contemporary furniture makers
of which the work of about 60 is on
display

WORDEN ARTS AND CRAFTS CENTRE
Worden Park, Leyland
Preston
Lancashire PR5 2DJ
Tel: 01772 455908
Fax: 01772 455908
Contact: Richard Blackburn
Established: 1984
Open: All year 10am to 4pm
Craftworkers on site: 9
Crafts: Stained glass, pyrography,
knitwear, ceramics, dried flowers,
bookbinder, blacksmith, stenciller, work
from North West Design Collective
Prices: Various
Note: Set in 157 acres of parkland, the
centre also has a theatre, 17th century
ice-house, miniature railway, maze,
garden for the blind, an arboretum,
children's play area, picnic area and
coffee shop.

THE POTTERS BARN
Roughwood Lane
Hassall Green
Sandbach
Cheshire CW1 0XX
Tel: 01270 884080
Contact: Andrew Pollard
Open: 9.30am to 6pm Monday to
Saturday. 1pm to 5pm Sunday
Craftworkers on site: 2
Crafts: Hand thrown domestic stoneware
and garden terracotta
Prices: 50p to £90. Access/Visa/
MasterCard/Eurocard accepted
Note: Canal and tearoom close by.

THE CRAFT GALLERY
114 Station Road
Ainsdale
Southport
Lancashire PR8 3PN
Tel: 01704 571421 or 01704 576032
Fax: 01704 570517
Contact: David and Janet Foster

Established: 1989
Open: 10am to 5.30pm Monday to
Friday, 10am to 4.30pm on Saturday
Stock: Watercolours, oils, pencil, ink
paintings. Sporting paintings a speciality.
Key rings, badges (die cast metal hand
cut, hand painted) as well as crafts of all
types
Prices: From £2 up to £50. No credit
cards accepted
Note: Graphic design also undertaken

CATHERINE SMITH CERAMICS
89 Scarisbrick New Road
Southport
Merseyside PR8 6LR
Tel: 01704 536048
Contact: Catherine Smith
Established: 1995
Open: Please telephone for details
Craftworkers on site: 1
Crafts: Handpainted porcelain china and
tiles. Dinner services, giftware, com-
memorative ware, bathrooms and
kitchens
Prices: £3 to £500. Credit cards not
accepted
Note: Classes held. Commissions welcome

RACEWOOD LTD
15A Park Road
Tarporley
Cheshire CW6 0AN
Tel: 01829 732006
Fax: 01829 732006
Contact: Bill Greenwood
Established: 1980
Open: 9am to 5pm Saturday
Craftworkers on site: 2
Crafts: Rocking horses, racehorse
simulators, trotting, cantering machines
Prices: £880 to £9,250. Credit cards not
accepted
Note: Rocking horses can be seen being
built and repaired

TODMORDEN CRAFT CENTRE
Lever Street
Todmorden
Lancashire OL14 5QF
Tel: 01706 818170
Contact: Mrs. M. Ford
Established: 1991
Open: 9.30am to 5pm all year except
Christmas Day and Boxing Day
Stock: Craft items made by local people,
needlework, soft toys, pottery,

woodturning, woollen rugs, jewellery
Speciality: Watercolour paintings and
picture framing
Prices: £1 to £150. Credit cards not
accepted
Note: Teashop with home made cakes
open all year

LUNAR SEA
1A Thomas Street
Widnes
Cheshire WA8 7SG
Tel: 0151 420 4061
Fax: 0151 420 7886
Contact: Sue France
Established: 1994
Open: 9.30am to 5pm Monday and
Tuesday. Closed Wednesday. 9.30am to
5.30pm Thursday, Friday and Saturday
Stock: Clothes, jewellery, crafts, textiles
from India, Bali, Thailand, Africa,
Equador, Nepal etc. Smoking parapherna-
lia and esoteric articles

KATY'S COFFEE & CRAFT SHOP
228 Town Lane, Bebington
Wirral
Merseyside L63 8LG
Tel: 0151 334 7823
Fax: 0151 639 1322
Contact: Miss L. Jones and Mrs. J. Jones
Established: 1995
Open: 9.30am to 5.30pm Monday to
Saturday
Stock: Country gifts, dried flower
arrangements, pottery, wooden pictures
Prices: £2.99 to £40
Note: Coffee shop

WIRRAL ROCKING HORSES
364 Hoylake Road, Moreton
Wirral
Merseyside L46 6DF
Tel: 0151 606 1177
Fax: 0151 606 1177
Contact: Stephen McGreal
Established: 1980
Open: 9am to 5pm Monday to Friday.
Close at 1pm on Saturday. Closed Sunday
and Bank Holidays. Please telephone first
if travelling a long distance
Crafts: Traditional hand carved rocking
horses, accessories and fittings for
restoration or making rocking horses.
Dolls houses
Prices: £399 to £1499. No credit cards
Note: Crafts Council listed

WOOD DESIGN WORKSHOPS
The Acre
Dappers Lane
Angmering
West Sussex
BN16 4EN
Tel: 01903 776010
Contact: Brendan Devitt-Spooner
Established: 1987
Open: 9am to 5.30pm Monday to Friday.
Weekends by appointment
Craftworkers on site: 2
Crafts: Contemporary furniture
Prices: £250 to £20,000-£30,000. Credit
cards not accepted

AMBERLEY MUSEUM
Amberley
Arundel
West Sussex
BN18 9LT
Tel: 01798 831370
Fax: 01798 831831
Contact: Robert Taylor
Established: 1979
Open: Wednesdays to Sundays mid-
March to the end of October plus Bank
Holiday Mondays and daily during local
school holidays
Craftworkers on site: up to 6
Crafts: Blacksmith, potter, boatbuilder,
clay pipe maker, broom maker, signwriter
and others
Prices: £1 to £10,000.
Note: Working museum of local industry
including major narrow gauge railway,
vintage open-top bus rides, cafe and other
visitor facilities. Disabled access

ARUNDEL GLASS STUDIO
Tarrant Square
Arundel
West Sussex
BN18 9DE
Tel: 01903 883597
Contact: Jacques Ruijterman
Established: 1979
Open: 9.15am to 5pm. Closed Sunday
and Monday
Craftworkers on site: 1
Crafts: Studio glass and engraved glass.
Quality crystal glass
Prices: £15 to £600. Credit cards not
accepted
Note: Engraving glass workshop held

FLOSSEY'S OF ARUNDEL
4 Tarrant Street
Arundel
West Sussex BN18 9DG
Tel: 01903 883073
Contact: Mrs. Irene Hall
Established: 1990
Craftworkers on site: 1
Open: 10.30am to 5pm Tuesday to
Friday. Saturday and Sunday all year
round 10.30am to 5.30pm
Craft/Stock: Porcelain dolls, arrange-
ments, painted and stencilled furniture,
pottery and hand crafts etc.
Prices: £1.50 to £350. Access/Visa/
MasterCard
Note: Items made on premises. Personal
service given. Situated opposite tearooms

SLINDON POTTERY
Top Road, Slindon
near **Arundel**
West Sussex BN18 0RP
Tel: 01243 814534
Contact: Janet Upton
Established: 1976
Open: Daily 9.30am to 11.30am and
1.30pm to 5.30pm. Sunday open from
10.30am. Closed Tuesday am
Craftworkers on site: 1
Crafts/stock: Thrown stoneware pottery
and clocks. Watercolours, turned wood,
jewellery, decorated hats, cards & candles
Prices: 50p to £95. Credit cards not
accepted
Note: Work can be seen in progress at
most times. Demonstrations by arrangement. Slindon is a National Trust village
on the edge of the South Downs with
beautiful views and walks

EVEGATE CRAFT CENTRE
Station Road, Smeeth
Ashford
Kent TN25 6SX
Tel: 01303 812334
Contact: Eric Jeanes
Open: Daily except Mondays but
including Bank Holidays
Crafts/stock: Pottery, woodturning,
silversmith, gift shop, art gallery, picture
framing, antiques, fudge, children's
clothes, photographic studio
Note: Restaurant, nature trails, rural
museum also on site. Disabled access

ALDERSHAW CRAFTS AND TILES
Aldershaw Farm, Kent Street
Sedlescombe
Battle
East Sussex TN33 0SD
Tel: 01424 754192
Fax: 01424 751462
Contact: Miss J. Lavis and Mr. J.R.
Williams
Established: 1983
Open: 9am to 5pm and weekends by
appointment
Craftworkers on site: 7
Crafts/stock: Handmade clay peg tiles,
floor tiles and special bricks. Stone statues
and garden urns and pots. Handpainted
wooden trays, boxes, children's toy
boxes, oil paintings on panel cards,
miniature watercolours on ivorine etc.
Prices: £5 to £500. Credit cards not
accepted
Note: Craft demonstrations, farm walk
and carp fishing. Disabled access

L'ARCHE WORKSHOP THE CANDLE FACTORY
11A Station Road
Bognor Regis
West Sussex PO21 2HS
Tel: 01243 821201
Contact: Mrs. Lindsay Wood
Established: 1987
Craftworkers on site: 2 +
Open: 9.30am to 4.30pm Monday to
Friday (closed weekends)
Crafts/Stock: Candles, rugs, necklaces,
greeting cards
Prices: £1 to £30. Credit cards not
accepted
Note: Visitors can see how candles are
made – also cards, beads and weaving.

HUGO BARCLAY
7 East Street
Brighton
East Sussex BN1 1HP
Tel: 01273 321694
Fax: 01273 725959
E-mail: hbarclay@cix.compulink.co.uk
Contact: Hugo Barclay/Lesley Randall
Established:1974
Open:10am to 1pm and 2pm to 5.30pm
Monday to Saturday
Stock: Glass, jewellery, ceramics, original
prints, wood, handmade cards
Prices: £10 to £2,000. Visa/Access/Amex/
Switch/MasterCard/Eurocard/Delta

Note: Two to four major exhibitions each
year. Mailing list by invitation. Member
Independent Craft Galleries Association.
Crafts Council Selected List of Shops &
Galleries. Disabled access to ground floor

SEAHORSE ARTISTIC METALWORK
17 Freshfield Place
Kemp Town
Brighton
East Sussex BN2 2BN
Tel: 01273 389297
Mobile: 0585 653046
Contact: Jane Killick
Established: 1993
Open: Monday to Friday 10.30am to
4.30pm
Craftworkers on site: 2
Crafts: Candleholders, mirrors, sconces
Prices: £1.50 to £50
Note: Small workshop in beautiful
countryside

CANTERBURY POTTERY
38A Burgate
Canterbury
Kent CT1 2HW
Tel: 01227 452608
Contact: Richard Chapman
Established: 1964
Open: Daily 9.30am to 6pm. Sunday
10.30am to 5.30pm
Craftworkers on site: 3
Crafts: Hand thrown domestic stoneware
pottery
Prices: £1 to £50. Visa/MasterCard
Note: Situated in centre of Canterbury,
opposite Cathedral, with its many varied
attractions and excellent shopping

THE GIBBS GALLERY AT CANTERBURY
53 Palace Street
Canterbury
Kent
CT1 2DY
Tel: 01227 763863
Contact: George and Dordie Gibbs
Established: 1991
Open: 10am to 5.30pm Monday to
Saturday. Closed Mondays Jan and Feb
Stock: Studio ceramics and sculpture in
various media
Prices: £10 to £1,500. All major credit
cards accepted
Note: Four major selling exhibitions each
year. Located 100 yards from Cathedral
entrance

MEDWAY CRAFT CENTRE
294-296 High Street
Chatham
Kent ME4 4NR
Tel: 01634 847809
Contact: Terry Webb
Established: 1977
Open: 9am to 5.30pm six days a week
Stock: Needlecraft, tapestries, cottons, wool, art material, picture framing, candlemaking, plaster casting, bead work, rubber stamping
Prices: 5p to £50

ORDNANCE MEWS CRAFT WORKSHOPS
Ordnance Mews
Historic Dockyard
Chatham
Kent ME4 4TE
Tel: 01634 830404
Contact: Robert Bright
Established: 1987
Open: Workshops open all year. Dockyard daily April to October 10am to 5pm. February, March and November Wednesday, Saturday and Sunday 10am to 4pm
Crafts/stock: Stained glass, model boats, furniture, ironwork, foundry, wood carver, jewellery, puppets
Prices: Various. All major credit cards accepted
Note: Visitors can see craftworkers at work in their studios. Many items for sale directly from makers. Situated in the living museum of Chatham Historic Dockyard. No entry charge to dockyard if visiting craft centre only

BRIDGE COTTAGE WORKSHOP
1 Bridge Cottage
Colworth
Chichester
West Sussex PO20 6DT
Tel: 01243 539623
Contact: Malcolm Stamp
Established: 1992
Open: 9am to 6pm but please telephone first
Craftworkers on site: 1
Crafts: Windsor chairs and country furniture
Prices: £50 to £500. Visa/Access
Note: Commissions undertaken and repairs and restoration of furniture are carried out

CRAFTWORK GALLERY
18 Sadlers Walk, East Street
Chichester
West Sussex PO19 1HQ
Tel: 01243 532588
Contact: Michael Pryke
Established: 1986
Open: 10am to 5pm Monday to Saturday
Stock: Wood, glass, pottery, silk scarves, cards, etchings, prints, watercolours, wall hangings, embroidery pictures
Prices: £1 to £150. Visa/Masterard
Note: Teashop adjacent to premises. Disabled access. The gallery is a contact point for the Guild of Sussex Craftsmen

THE BANK GALLERY
73-75 High Street
Chobham
Surrey GU24 8AF
Tel: 01276 857369
Open: Tuesday to Saturday 10am to 5pm. Closed Bank Holidays
Stock: Domestic pottery, studio ceramics, jewellery, wood, metal, textiles, glass, basketry
Note: Wheelchair access with help

RONALD EVERSON
Willowdene
Blundel Lane, Stoke D'Abernon
Cobham
Surrey KT11 2SP
Tel: 01372 842313
Contact: R. Everson
Established: 1971
Open: 10am to 5pm but please telephone first
Crafts: Stained glass, jewellery, porcelain, bone china interior decorative and functional products
Prices: £2 to £250
Note: Crafts Council listed. Associate of Royal College of Art

JILL PRYKE POTTERY
Unit 2
The Turner Dumbrell Workshops
Ditchling
East Sussex BN6 9TD
Tel: 01273 845246
Contact: Jill Pryke
Established: 1975
Open: 10am to 5pm Tuesday to Saturday
Craftworkers on site: 1
Crafts/stock: Pottery
Prices: £5 to £50. Credit cards accepted

THE CRAFTSMAN GALLERY
8 High Street
Ditchling
East Sussex BN6 8TA
Tel: 01273 845246
Contact: Jill Pryke
Established: 1975
Open: 10am to 5pm Monday to
Saturday. Closed Wednesdays
Stock: Ceramics, turned wood, jewellery
and some textiles as well as prints and
watercolours interpreting Sussex landscape.
Note: Set in the village of Ditchling with
its famous craft history. Local museum,
tearooms and pubs nearby.

TURNER DUMBRELL WORKSHOPS
North End
Ditchling
East Sussex BN6 8TD
Tel: 01273 846338
Fax: 01273 846684
Contact: Anton Pruden and Rebecca
Smith
Established: 1989
Open: 10am to 6pm Tuesday to Saturday
Craftworkers on site: 7
Crafts: Pruden & Smith Silversmiths,
Kevin Hutson Woodturner, Treeline
Cabinetmakers, Perry Lancaster, carved
wood. Jill Pryke, pottery, Gemma
Quinton, designer clothing
Prices: £20 to £1,000. Access/Visa/Amex
Note: Set in a complex of workshops
Limited disabled access

THE GREAT BRITISH CRAFT COMPANY
Old Kings Head Court
Dorking
Surrey RH4 1AR
Tel: 01306 889355
Established: 1993
Open: 9.30am to 5.30pm every day
except Sunday
Stock: Ceramics, glass, wood
Note: Regular themed exhibitions held.
Also at 139 High Street, Banstead

THE BANANA TREE
The Enterprise Centre, Station Parade
Eastbourne
East Sussex BN21 1BE
Tel: 01323 647713
Contact: Lucy Pike
Established: 1990
Open: 9.30am to 5pm Monday to Saturday
Stock: Unusual gifts from the Far East.

Soft furnishings from India and China.
Candles, incense and oils, recycled
glassware, jewellery
Speciality: Large range of throw-overs
Prices: 25p to £500. Access/Visa/
MasterCard/Eurocard/Amex
Note: Situated in arcade with 50 other
specialist shops

ALAN LOCK FRAMING
36 Northend Road
Erith
Kent DA8 3QE
Tel: 01322 342975
Contact: Alan Lock
Established: 1989
Open: Most days but please phone first
Craftworkers on site: 1
Crafts: Bespoke picture framing, mount
cutting for a wide variety of displays
Prices: £6 to £1,000

MANOR FARM CRAFT CENTRE
Wood Lane, Seale
near **Farnham**
Surrey GU10 1HR
Tel: 01252 783661
Contact: Janice Midgley
Established: 1982
Open: 11am to 5pm daily. Sundays 2pm
to 5pm. Closed Mondays except Bank
Holidays
Stock: Pottery, glass engraving and textile
art, calligraphy and stone carving, soft
furnishing, machine knitting. General
craft shop for gifts and hand-made
greetings cards
Prices: £1 to £150. Credit cards not
accepted
Note: Teashop, working crafts people,
lovely church next door and garden
centre close by. Situated in picturesque
old farm buildings. Close to North Downs
Way amid beautiful walking country

NEW ASHGATE GALLERY
Wagon Yard
Farnham
Surrey GU9 7PS
Tel: 01252 713208
Fax: 01252 737398
Contact: Susan Szabanowicz/Joanne
Barber
Established: 1976
Open: 10am to 5pm Monday to Saturday
Stock: High quality ceramics, sculpture,
jewellery, glass, metalwork, furniture as

well as textiles all by professional makers
Prices: £30 to £3,500. Visa/Access/Switch
Note: Crafts Council approved. Craft shop
and changing exhibitions featuring the
work of established artists and makers
and talented newcomers

APPLE CRAFT CENTRE
Macknade, Selling Road
Faversham
Kent ME13 8XF
Tel: 01795 590504
Contact: Sue and Ian Sandford
Established: 1992
Open: 9.30am to 5.30am. Coffee shop/
restaurant 10am to 5pm. Closed 25, 26,
27 December
Stock: Quality handmade craftwork made
by Kent craftworkers
Prices: From under £1 upwards
Note: Woodturning tuition and supplies.
Beautiful coffee shop and restaurant
serving homemade food. Craft workshops
and demonstrations on site. Programme
of craft related exhibitions throughout the
year. Located on A2 at Faversham one
mile from M2 junction 7. Free car park,
farmshop and garden centre adjacent

RICKY HOLDSTOCK – SEAT WEAVER
Hillside Cottage
The Forstal, Hernhill
Faversham
Kent ME13 9JQ
Tel: 01227 751204
Fax: 01227 751204
Contact: Mr. M.R.L. Holdstock
Established: 1975
Open: 9am to 5pm daily, but appoint-
ments appreciated
Crafts: Chair seat weaving in rush, cane,
cord etc.
Prices: £15 to £2,000. No credit cards
Note: Working craft shop. Visitors
welcome to watch work in progress but
advance notice recommended as parking
limited. Member of Guild of Master
Craftsmen, Basketmakers Association and
Conservation Register listed

SURREY GUILD CRAFT SHOP &
GALLERY
1 Moushill Lane
Milford
Godalming
Surrey
GU8 5BH

Tel: 01483 424769
Contact: Zoe O'Brien and Helena Greig
Established: 1993
Open: 10.30am to 5pm seven days a
week. Closed Christmas and Boxing Day
Stock: Jewellery, ceramics, woodturning,
furniture, weaving, silk painting, knitting,
calligraphy, quilting, stencil and
découpage
Prices: 30p to £600. Access/Visa/
Eurocard
Note: Weekly exhibition by Craftsman of
the Week plus occasional visiting
exhibitions. Weaving workshop open to
visitors when in use. Courses held. All
work is by exhibiting members of Surrey
Guild of Craftsmen. Work can be bought
direct from maker at craftsmen's own
prices

FLINT BARN
Old Loom Mill
Hersham Road
Hailsham
East Sussex
TN39 3UW
Tel: 01424 845301
Contact: Mrs. B. Butler
Established: 1991
Open: 9am to 5pm Monday to Saturday.
10am to 4pm Sunday
Stock: Spinning and weaving, cross stitch,
painted cards, wood turning, hats and
bags, patchwork, china painting, dried
flower pictures
Prices: £1 to £48. Visa/Access
Note: Demonstrations by arrangement,
teashop, children's area, large barn
selling fabric and wool – mostly end of
lines

HENFIELD WOODCRAFT
Harwoods Farm, West End Lane
Henfield
West Sussex BN5 9RF
Tel: 01273 492820
Contact: P.D. Spear
Established: 1966
Open: By appointment
Craftworkers on site: 1
Crafts: Salad bowls and platters in English
hardwoods. Thin walled goblets
Prices: £1.50 to £1,000. Credit cards not
accepted
Note: All work to order only. Normal
delivery time 2 years. Small workshop.
No access for disabled to workshop

THE TRUGGERY
Coopers Croft
Herstmonceux
East Sussex BN27 1QL
Tel: 01323 832314
Fax: 01323 832314
Contact: Sarah Page
Established: 1899
Open: Tuesday to Saturday 10am to 5.30pm
Craftworkers on site: 2
Crafts: Sussex trug baskets in willow/chestnut. Also walking sticks and English basketware
Note: Public may visit workshop to see trugs being made. Small workshop so not too many visitors at once.

THOMAS SMITH'S TRUG SHOP
Hailsham Road
Herstmonceux
East Sussex
BN27 4LH
Tel: 01323 832137
Fax: 01323 833801
Contact: Robin Tuppen
Established: 1829
Open: 7.30am to 5pm Monday to Friday. 9am to 4pm Saturday
Craftworkers on site: 16
Crafts: Sussex trug baskets, walking sticks, corn dollies, wooden fruit
Note: Purchasers of trugs may visit workshops. Guided tours arranged for groups of 10 or more (£2 per person)

DIAMOND FIBRES LTD.
Diamonds Farm
Horam
East Sussex
TN21 0HF
Tel: 01435 812414
Established: 1988
Open: Thursday to Saturday 2.30pm to 5.30pm Easter to 31st October. Other times by appointment
Crafts: Individually designed handknitted garments in naturally coloured wool. Naturally coloured yarn
Prices: Up to £100. Credit cards not accepted

DAISY'S CRAFTS
The Cottage
Horsebridge
East Sussex
BN27 4DJ

Tel: 01323 440933
Contact: J.A. Best
Open: Tuesday to Saturday 10am to 5pm
Stock: Various craft items
Note: Courses held in stencilling, paint effects, silk painting, renovating junk furniture and designing and cutting out own stencils

FIRE AND IRON GALLERY
Rowhurst Forge
Oxshott Road
Leatherhead
Surrey
KT22 0EN
Tel: 01372 375148
Fax: 01372 375148
Contact: Lucy Quinnell
Established: 1984
Open: 10am to 5pm Monday to Saturday
Craft/Stock: Metalwork (jewellery, furniture, sculpture, garden pieces, fire furniture, etc.)
Prices: £8 to £8,000. Visa/Access/Barclaycard
Note: Set in picturesque grounds of 1450 farmhouse. Lively programme of exhibitions and forging demonstrations. Selected for quality by the Crafts Council

CHARLESTON SHOP/GALLERY
Charleston Farmhouse
near Firle
Lewes
East Sussex BN8 6LL
Tel: 01323 811626
Fax: 01323 811628
Contact: Alistair Upton
Established: 1987
Open: Shop Easter or April 1st to October: Wednesday to Sunday and Bank Holidays 2pm to 6pm. November and December Saturday and Sunday only, 2pm to 5pm. Mid-January to March by appointment only. House closes at 5pm
Stock: Unique range of high quality handmade hand-decorated ceramics, textiles and furniture. Changing exhibition programme
Prices: From £10 to £100. Most major cards accepted and Switch
Note: This outlet is adjacent to the "Bloomsbury House" which displays many pictures, artifacts and memorabilia relating to the famous Bloomsbury group of intellectuals. Entrance to the house is £5 and concessions £3

GUILD OF SUSSEX CRAFTSMEN SHOP
Bentley Wildfowl & Motor Museum
Halland
Lewes
East Sussex
BN8 5AF
Tel: 01825 840573
Contact: Barry Sutherland
Established: 1996
Open: mid March to end October 10.30
am to 4.30 pm. Weekends in November,
February & early March 10.30am to 4pm
Stock: From Guild Members: textiles,
silver, glass, ceramics, metals, jewellery,
engraving, weaving, batik and wood
Prices: From under £5 to over £1000.
Credit cards not taken at present
Note: Situated on the beautiful Bentley
estate with access to the museum and
other visitor activities

JONATHAN SWAN/THE WORKSHOP
164 High Street
Lewes
East Sussex
BN7 1XU
Tel: 01273 474207
Contact: Susanna Chaplin
Established: 1981
Open: 9am to 5.30pm Monday to
Saturday
Crafts/stock: Jewellery, mostly silver and
contemporary.
Prices: £5 to £250. Visa/MasterCard
Note: Commissions undertaken. Crafts
Council selected shop for quality. Evening
courses held in jewellery making three
nights a week

MARY POTTER STUDIO
Shortgate Lane
Laughton
Lewes
East Sussex
BN8 6DE
Tel: 01825 84 438
Contact: Mary Potter
Established: 1975
Open: Usual shop hours and weekends.
Closed during May and part of Septem-
ber. Advisable to phone first
Craftworkers on site: 2
Crafts: Batik pictures, collage pictures,
cards
Note: Sketch books, photographs of
sources, methods can be seen. Located in
pleasant rural surroundings

THE OLD STABLES CRAFT CENTRE
The Old Stables, Market Lane
Lewes
East Sussex BN7 2NT
Tel: 01273 475433
Contact: Mrs. Karen Marsh
Established: 1992
Open: 9.30am to 5.30pm Monday to
Saturday. 10.30am to 4.30pm Bank
Holidays (closed May Bank Holiday)
Crafts: Stained glass – lampshades,
mirrors, display cases, window hangers
and terrariums. Silver and pewter giftware
Prices: £3 to £300. Visa/Mastercard
Note: Other crafts: millefiori jewellery,
quilted cushions and quilts, local prints,
cards and many other crafts. Tea and
coffee shop selling specialist foods.
Parking and toilets. Access for disabled

KENT POTTERS ASSOCIATION/ GALLERY
70 Bank Street
Maidstone
Kent ME14 1SN
Tel: 01622 863554
Contact: Janet Jackson
Established: Gallery 1994. (KPA 1974)
Open: Tuesday to Saturday 10am to 5pm
Craftworkers on site: None working but
member of association at the Gallery .
Stock: Ceramics by members of the KPA
throughout Kent. Fullest possible range of
ceramics from decorative to functional in
stoneware, porcelain and terracotta
Prices: £1.50 to £200
Note: There is also a smaller gallery
holding monthly exhibitions of work
other than ceramics

RAYMENT & HULL SILVERSMITHS
Minster Museum, Bedlam Court Lane
Minster
Thanet, Kent CT12 4HQ
Tel: 01843 822619
Fax: 01843 821635
Contact: Neil Rayment and Robert Hull
Established: 1990
Open: 9am to 5pm Monday to Friday
(also at Sidney Cooper Centre, Canter-
bury, every Saturday 9am to 5pm)
Crafts: Small stock of silver jewellery but
we specialize in making to order
jewellery and silverware; also repairs
Prices: £2 to many thousands
Note: Museum has teashop, other crafts
on show and farm animals. Picnic area

HANNAH PESCHAR GALLERY
Black and White Cottage
Standon Lane
Ockley
Surrey RH5 5QR
Tel: 01306 627269
Fax: 01306 627662
Contact: Hannah Peschar
Open: Friday to Saturday 11am to 6pm.
Sunday 2pm to 5pm. Closed Monday,
Tuesday and Wednesday. Thursday by
appointment
Stock: Sculpture and studio ceramics,
wood
Note: Sculpture and ceramics displayed
in controlled wilderness of water garden.
Entrance to garden £4, children £2,
concessions £3. Guided tours with
refreshments by appointment

GRAFFHAM WEAVERS
Glasses Barn
Graffham
Petworth
West Sussex   GU28 0PU
Tel: 01798 867348
Contact: Miss Barbara Mullins
Established: 1952
Craftworkers on site: Yes
Open: Open most days but please phone
first
Crafts: Floor rugs, woven cushions, bags,
garments
Prices: £150 to £400. Credit cards not
accepted
Note: Short courses arranged. Two
exhibitions held each year in May and
November

GLYNLEIGH STUDIO
Peelings Lane
Westham
Pevensey
East Sussex   BN24 5HE
Tel: 01323 763456
Contact: Sam Fanaroff
Established: 1958
Open: 9am to 5.30pm Monday to Friday
Crafts: Copper and brass bowls, vases,
trays, candlesticks, shepherds' crooks,
lamps, firescreens. Ecclesiastical work.
Restoration and repair to all non-ferrous
metals
Note: Commissions undertaken for
private individuals and institutions. Many
attractions nearby including sea (2 miles),
castle, church walks, antique shops, pubs

KATHI'S HAND-PAINTED CHINA & EPOKA – TRADITIONAL LEATHER BOXES & TRUNKS
9 Worple Way
Richmond
Surrey
TW10 6DG
Tel: 0181 940 5994
Fax: 0181 940 5994
Contact: Kathi Rindlisbacher and Iwona
Pietraszewska
Established: 1984 and 1996
Open: 10am to 1pm or by appointment
Monday to Saturday
Craftworkers on site: 2
Crafts: Unusual hand-painted china and
porcelain: tablelamps, vases, plates, etc.
Hand-crafted leather boxes which may be
personalised
Prices: £15 to £600
Note: Joint open evening every Novem-
ber. Individually tailored courses in china
painting (maximum 3 people)
Note: Commissions and trade enquiries
welcome

STUDIO ONE
1 Victoria Street
Rochester
Kent
ME1 1XY
Tel: 01634 811469
Contact: E. Falcke, C. Murrell, A.
Wheeldon
Open: Most weekdays 10.30am to
4.30pm but advisable to make appoint-
ment
Crafts: Ceramics and silver and gold
jewellery
Prices: £8 to £150 approx. Credit cards
not accepted
Note: Commissions undertaken. Located
in historic Rochester-upon-Medway with
castle and cathedral plus many Dicken-
sian features

CHAPEL GALLERY
29 Park Road
Rottingdean
East Sussex BN2 7HR
Tel: 01273 308699
Fax: 01273 300819
Contact: Mike and Sue Jerrome
Established: 1992
Open: 7 days a week. Weekdays 9.30am
to 5pm. Sundays and Bank Holidays
11am to 5pm

Stock: Original paintings by local artists, prints, limited editions, gift stationery, art and craft materials. Picture framing available on the premises. Large selection of books
Prices: 50p to £1200. Access/Visa/Mastercard/Switch
Note: Many art and craft demonstrations in the chapel. Painting courses, calligraphy and screen printing courses held. Jazz workshop every Monday evening. The gallery can be approached through the shop as well as 29 Park Road

JERROMES
45 High Street
Rottingdean
East Sussex
BN2 7HR
Tel: 01273 308699
Fax: 01273 300819
See entry for CHAPEL GALLERY above for details of Stock/Prices etc

THE GALLERY
6 Nevill Road
Rottingdean
East Sussex BN2 7HG
Tel: 01273 306081
Contact: Ms. Lawson and Ms. Blakelock
Established: 1994
Open: Six days 9am to 5pm. Closed all day Wednesday and 1pm to 2pm every day
Stock: Hand crafted work – high quality pottery, hand blown glass, stained glass, sculpture, pressed flower crafts, mosaic etc. Unusual greeting cards (many hand made)
Speciality: Original watercolours and oil paintings
Prices: £5 to £500. Visa/MasterCard/Eurocard
Note: There is a tearoom serving hand made cakes

RYE ART GALLERY
Easton Room
107 High Street
Rye
East Sussex
TN31 7JE
Tel: 01797 223218
Contact: Miranda Leonard
Established: 1965
Open: 10.30am to 5pm daily all year.

Closed only for re-hanging
Stock: Jewellery, textiles, ceramics, metal and glass
Speciality: Two high quality contemporary craft shows a year
Prices: £20 to £2,000. Visa/Access
Note: Adjacent sister gallery Stormont Studio. Historical and contemporary fine art exhibitions programme. Permanent collection. Burra memorabilia cabinet. Active workshop programme. Guided tour by curator by arrangement

SEAL CHART STUDIO
Chart Farm Craftwork Shops
Seal Chart
Sevenoaks
Kent
TN15 0ES
Tel: 01732 462148
Contact: Lawrence Jenkins
Established: 1986
Open: Monday 10am to 5pm and Thursday 10am to 5pm or by appointment
Craftworkers on site: 12
Crafts: Etching and intaglio prints. Paintings
Prices: £10 to £1000. Credit cards not accepted
Note: This is part of a craft workshop centre and there are other craft units in the Farm Yard including dried flowers, cabinet making (see below) etc. that can be visited

T. CONSTABLE
Workshop 6
Chart Farm
Seal Chart
Sevenoaks
Kent TN15 0ES
Tel: 01732 763263
Contact: Tim Constable
Established: 1982
Open: Monday, Tuesday, Wednesday, Friday 9am to 5pm
Craftworkers on site: 1
Crafts: Furniture and other contemporary items in wood
Prices: £5 to £3,000. Credit cards not accepted
Note: Part of above mentioned craft workshop centre. Working workshop with small range of items on display. Commissions accepted. Please ring for appointment

PICTURES PLUS ARTS AND CRAFT CENTRE
185 High Street
Sheerness
Kent
ME12 1UJ
Tel: 01795 583503
Contact: Jean White and J. Adams
Established: 1993
Open: 9am to 5pm
Crafts/stock: All handmade local crafts. Pictures, arts and crafts of all kinds, craft and art materials, gifts
Prices: £2 to £350. Barclaycard/Visa/Connect
Note: Demonstrations by expert craft people, workshop courses in art, ceramics, dough making, patchwork etc.

SIDCUP ANTIQUE & CRAFT CENTRE
Elm Parade
Main Road
Sidcup
Kent
DA14 6NF
Tel: 0181 309 5032
Fax: 0181 308 0748
Contact: Maurice Tripp
Established: 1993
Open: 10am to 5pm 7 days a week (closed Christmas Day, Boxing Day and New Year's Day)
Stock: Ceramics, jewellery, dolls, dolls' houses, toys, teddy bears, lace, knitwear, hats, candles, dried and silk flowers, eggcraft, glass engraving, stained glass, wood carving, marquetry, soft furnishing, military models, craft supplies, cards, painting, portraits, beads, tapestry, doughcraft etc.
Prices: £1 to £175. Access/Visa/Switch/Amex
Note: Over 90 dealers and craftspeople. Coffee shop and easy parking

P.G. BLEAZEY WOODWIND INSTRUMENTS
63 Middletune Avenue
Milton Regis
Sittingbourne
Kent
Tel: 01795 470149
Fax: 01795 470149
Contact: Phil Bleazey
Established: 1992
Open: Most days but please 'phone first
Crafts: Flutes, recorders, whistles and early percussion from medieval to modern folk
Speciality: Instruments for historical re-enactments
Prices: £50 to £250. Credit cards not accepted
Note: Short hands-on courses in instrument making techniques. Short talks on instrument care. Musicians are always welcome to play instruments

CATHIE BEAZLEY - DECORATIVE ROPEWORK
see address and telephone details above
Established: 1996
Craft: wall hangings, mats, jewellery all hand crafted from rope including items for historical re-enactments
Prices: £5 to £50

CAROL EDWARDS – LEATHERCRAFT & FLETCHER
Craft Cottage
Bookham Lodge Stud
Cobham Road
Stoke D'Abernon
Surrey
KT11 3QG
Tel: 01932 865181
Contact: Carol Edwards
Craftworkers on site: 1
Established: 1982
Open: 2.30pm to 6.30pm Monday to Friday. 2pm to 5.30pm some Saturdays
Crafts/Stock: Archery equipment (traditional), leather goods, trophies
Speciality: English longbows and wood arrows
Prices: £4 to £200. Credit cards not accepted
Note: Archery tuition. Corporate days

DAVID HAUGH FURNITURE
Unit 7
Noblesgate Farm and Yard
Bells Yew Green
near **Tunbridge Wells**
East Sussex
TN2 4SJ
Tel: 01892 750310
Fax: 01892 520598
Contact: David Haugh
Established: 1977
Open: 9am to 5pm Monday to Saturday
Craftworkers on site: 3
Crafts: Quality handmade furniture (built in and free standing) formed from

hardwoods, oak, cherry, walnut, beech etc
Prices: £5 to £5,000. Credit cards not
accepted
Note: Customers are welcome to watch
furniture being made. Most of work is
commissioned. Showroom next to
workshop has examples of work, timber
samples and photographs

BRIAR BANK SOFTWEAR

Briarbank, Highview Lane
Ridgewood
Uckfield
East Sussex
TN22 5SY
Tel: 01825 767757
Fax: 01825 767757
Contact: Mrs. Ann Dishman
Established: 1994
Open: Most days but please 'phone first
Craftworkers on site: 1
Crafts: Handspun yarns in luxury fibres,
e.g. homegrown angora fibre with silk.
Handspun garments, painted silk scarves
and wraps
Prices: £2.50 to £60. Credit cards not
accepted
Note: Some handweaving and spinning
demonstrations. Also silk painting and
spinning tuition. Visitors can see angora
rabbits – the source of the garments and
yarn

LIVING CERAMICS

Bird-in-Eye Farm
Framfield Road
Uckfield
East Sussex
TN22 5HA
Tel: 01825 890163
Contact: Clare McFarlane
Established: 1984
Open: 9am to 5pm Monday to Friday.
9am to 1pm Saturday
Craftworkers on site: 1 1/2
Crafts: Hand painted semi-porcelain cats,
pigs, chickens, frogs etc. made from
moulds of my own original designs
Prices: £2.80 to £40. Credit cards not
accepted

SHARLANDS FARM POTTERY

Sharlands Farm
Blackboys
Uckfield
East Sussex
TN22 5HN

Tel: 01435 862652
Contact: Matthew Bayman
Established: 1985
Open: 10am to 5.30pm Monday to
Saturday. Sunday by appointment
Craftworkers on site: 1
Craft/stock: Ash glazed wood fired
stoneware. Wheel thrown domestic and
individual pots. Original gouache
paintings
Prices: £4 to £150. Credit cards not
accepted
Note: Adult evening and day classes.
Member of The Guild of Sussex
Craftsmen

WILDERNESS ROAD

Hadlow Down
near **Uckfield**
East Sussex
TN22 4HJ
Tel: 01825 830509
Fax: 01825 830977
Contact: Chris Yarrow
Established: 1985
Open: 10am to 5pm daily
Craftworkers on site: 2
Crafts: Wooden garden furniture and
rustic garden work. Trugs, hazel hurdles,
turned wooden bowls and pebbles
Prices: £1 to £300. Credit cards not
accepted
Note: Working woodland open to visitors
with trails, exhibition, teas, picnic area,
adventure playground. Much of raw
material grown on site. Commissions
taken for made-to-measure work.
Disabled access (sales area, workshop,
WCs)

MILL YARD CRAFT CENTRE

Swan Street
West Malling
Kent
ME19 6LP
Tel: 01732 845888
Fax: 01732 845888
Contact: R.L. and R.M.A. Lovegrove
Established: 1985
Craftworkers on site: approx 10
Open: 9.30am to 5pm Monday to
Saturday but times may vary for each
unit. Closed Sundays and Bank Holidays
Crafts/stock: Art gallery and framing,
artists' materials, ethnic gifts, découpage,
doll and teddy hospital, haberdashery/
stitchcrafts, greetings cards, designer

cakes and cake crafts, dried flower
workshop, goldsmith and jeweller,
Note: There are a number of units on this
site from which crafts are made. Classes
held in stitchcraft, découpage, cakecraft,
art. Tearoom, farm shop, spiritual healing
and seasonal tourist information

SHIRLEY'S DÉCOUPAGE STUDIO
Millyard Craft Centre (see above)
Swan Street
West Malling
Kent
ME19 6LP
Fax: 01732 872917
Contact: Shirley Hills
Established: 1992
Open: 9.30am to 5pm (closed Sundays).
Half day Wednesday. Closed all day
Monday
Stock: 3D découpage pictures and kits.
Gifts, greeting cards and 3D découpage
glue, varnish, scissors etc.
Prices: £5 to £100+. No credit cards at
present
Note: Classes are held in 3D découpage

PETER HACKFORD POTTERY
25 Northwood Road
Whitstable
Kent
CT5 2EU
Tel: 01227 274857
Contact: Peter Hackford
Established: 1978
Open: 9am to 5pm Monday to Thursday.
Please telephone before visiting
Crafts: Stoneware, porcelain, earthen-
ware, domestic and individual pieces
Prices: £3 to £60. Credit cards not
accepted
Note: Member of Kent Potters Association
and teacher at adult centres and schools

S O U T H E R N

J. & L. POOLE
38 Wootton Road
Abingdon
Oxfordshire
OX14 1JD
Tel: 01235 520338
Contact: M.J. Poole
Established: 1985
Open: By appointment only 7 days per week
Craftworkers on site: 1
Crafts: Silverware, silver and gold jewellery and silver cased clocks
Prices: £5 to £1,000. Credit cards not accepted
Note: Honorary life member of Oxfordshire Craft Guild. Studio located in private house. Easy parking

CANDOVER GALLERY
22 West Street
Alresford
near Winchester
Hampshire SO24 9AE
Tel: 01962 733200
Contact: Barbara Ling
Established: 1984
Open: Monday to Saturday 9.30am to 5.30pm
Crafts/stock: High quality studio ceramics, glass, jewellery as well as original paintings and prints
Prices: From around £25 to £2500. All major credit cards accepted plus Switch and Delta
Note: This high quality gallery is located in a 200 year old building in the historic town of Alresford

SOPHIE PATTINSON WALLHANGINGS
No. 3 Sunnyside Cottages
Church Street, Ropley
Alresford
Hampshire
SO24 0DP
Tel: 01962 772516
Contact: Sophie Pattinson
Established: 1987
Open: 9am to 6pm or by appointment
Craftworkers on site: 1
Crafts: Painted and woven wall hangings
Prices: £45 to £1,000. Credit cards not accepted
Note: Demonstrations of weaving. Set in rural countryside

FROYLE POTTERY
Lower Froyle
Alton
Hampshire GU34 4LL
Tel: 01420 23693
Fax: 01420 22797
Contact: Mrs. Candida Griffin
Established: 1983
Open: 9am to 5pm
Craftworkers on site: 6+
Crafts: Range of pottery and handmade ceramic tiles
Prices: £12 to £400. Credit cards not accepted

HAMPSHIRE AND BERKSHIRE GUILD OF CRAFTSMEN
The Gallery, New Farm Buildings
Lasham
near **Alton**
Hampshire GU34 5RY
Tel: 01256 381368
Contact: Hugo Egleston
Established: 1995
Craftworkers on site: 1
Open: Most days but please 'phone first
Craft/Stock: High quality jewellery, textiles, furniture, glass, ceramics as well as original prints
Prices: From £20 to £6000. Some credit cards accepted
Note: Regular exhibitions held. Commissions welcomed. This gallery is attached to a studio where Hugo Egleston makes fine furniture from English hardwoods

SELBORNE POTTERY
The Old Bakehouse
The Plestor, Selborne
near **Alton**
Hampshire GU34 3JQ
Tel: 01420 511413
Fax: 01420 511233
Contact: Robert Goldsmith
Established: 1985
Open: 9am to 6pm. 11am to 5pm Saturday and Sunday
Craftworkers on site: 3
Crafts: Hand-thrown and decorated domestic stoneware
Prices: £5 to £200. Mastercard/Visa
Note: Situated in beautiful cobbled courtyard behind the Selborne Gallery. Beautiful village, National Trust walks, tea rooms, Gilbert White Museum,'The Wakes'

THE CHILTERN BREWERY
Nash Lee Road, Terrick
Aylesbury
Buckinghamshire HP17 0TQ
Tel: 01296 613647
Fax: 01296 612419
Contact: Mr. and Mrs. R. Jenkinson
Established: 1980
Open: 12 noon Saturday Tour £3.
Essential to phone to check availability
Craftworkers on site: 5
Crafts/stock: Real ale brewery. Beer-
related produce i.e. beer mustards,
cheeses, malt marmalades etc.
Prices: £1.20 to £50. Credit cards
accepted
Note: Small beer museum

JACQUELINE PURTILL JEWELLERY
Unit B14 German Road
Bramley Green
near **Basingstoke**
Hampshire RG26 5BG
Tel: 01976 264840
Contact: Jacqueline Purtill
Established: 1994
Open: Saturday 10am to 4pm and other
times by appointment
Craftworkers on site: 1
Crafts: Wide range of designer and
handmade jewellery using silver, silver
wire and semi-precious stones
Prices: From £20 to £250

VIABLES CRAFT CENTRE
Harrow Way
Basingstoke
Hampshire RG22 4BJ
Tel: 01256 473634
Fax: 01256 58086
Contact: M Wright
Established: 1975
Open: 10am to 4pm Tuesday to Saturday.
Occasional Sundays but phone before
starting out
Craftworkers: 14
Crafts: Woodturning, felt making,
ceramics, cake decorations and accesso-
ries, gold and silver jewellery, picture
framing, silk flowers, beauty therapy,
glass engraving, metal engraving, stained
glass, bridal gowns
Prices: 50p to £500
Note: A workshop centre where a number
of craft workers work. New restaurant and
bar, summer exhibiton gallery, events
throughout the year. Miniature railway

EMMA LUSH CERAMIC ARTIST & DESIGNER
The Studio
Botley Mills
Botley
Hampshire SO30 2GB
Tel: 01489 782006
Contact: Emma Lush
Established: 1994
Open: Most days but only by appoint-
ment. Also sells from exhibitions
Craftworkers on site: 2
Crafts: Ceramic furniture, mirror candle
holders, candle sticks, bowls, urns,
mirrors, wine coolers, planters and tiles
Prices: £1.50 to £3,500. Visa/Access/
MasterCard/Eurocard/Switch/JCB

MIRIAM TROTH
125 Seafield Road
Southbourne
Bournemouth
Dorset BH6 3JL
Tel: 01202 427296
Contact: Miriam Troth
Established: 1985
Open: 2pm to 4pm Wednesdays. Please
telephone for details
Craftworkers on site: 1
Crafts: Jewellery and small sculpture.
Contemporary designs created from
found objects, silver, gold leaf and semi-
precious stones
Prices: £25 to £75. Credit cards not
accepted
Note: No disabled access

THE PAUL CREES COLLECTION
124 Alma Road
BOURNEMOUTH
Dorset
BH9 1AL
Tel: 01202 716786
Fax: 01202 716786
Contact: Peter Coe
Established: 1980
Open: 11am to 5pm
Craftworkers on site: 2
Crafts: Original limited edition wax
dolls
Prices: £300 to £2500. Amex/
Mastercard/Visa all accepted

WESTBOURNE CRAFT CENTRE
13 Seamoor Road
Westbourne
Bournemouth
Dorset BH4 9AA
Tel: 01202 766532
Fax: 01202 533090
Contact: Mrs. Jenny Compton
Established: 1995
Open: 9am to 5pm Monday to Saturday
Stock: Tapestry, cross-stitch, cushions and stools
Prices: £7 to £300. Credit cards not accepted

DRAGON GALLERY
19A Market Hill
Buckingham
Buckinghamshire MK18 1JX
Tel: 01296 730303
Contact: Elizabeth Linton
Established: 1984
Open: 9.30am to 4.45pm Monday to Saturday
Craftworkers on site: 1
Crafts/stock: Ceramics, doughcraft, jewellery, greeting cards
Prices: 99p to £75. Credit cards not accepted

CRAFTWORK
18 Sadlers Walk
East Street
Chichester
West Sussex PO19 1HQ
Tel: 01243 532588
Contact: Michael Pryke
Established: 1989
Open: 10am to 5pm Monday to Saturday
Stock: Wood, glass, stained glass, pottery, etchings, prints, watercolours, wall hangings, cards
Prices: £2 to £150. Visa/Mastercard
Note: Vice President of the Guild of Sussex Craftsmen

THE GLASS WORKSHOP
Kestrel Cottage
Spriggs Holly Lane, Spriggs Alley
near **Chinnor**
Oxfordshire OX9 4BU
Tel: 01494 483374
Contact: Gill Cox
Established: 1988
Open: Please ring for details
Craftworkers on site: 1 or 2
Crafts: Carved, engraved and cast multi-layered coloured glass. Specialist in 'Graal' glass
Prices: £5 to £1,500. No Credit cards
Note: Continuous demonstrations. Spriggs Alley is located in a remote area of outstanding natural beauty in the Chilterns, close to Junction 6 of M40. Surrounded by ramblers' routes and footpaths centred on the Ridgeway and the Icknield Way.

OXFORDSHIRE CRAFT GUILD SHOP
7 Goddards Lane
Chipping Norton
Oxfordshire OX7 5NP
Tel: 01608 641525
Contact: Valerie Newey
Established: 1994
Open: Tuesday to Saturday 10am to 5pm
Stock: Glass, pottery, pewter, textiles, jewellery, wood, silverware
Prices: £4.50 to £90. Credit cards not accepted
Note: Shop is staffed by makers. A co-operative of 22 all selected to be Guild members. Exhibitions displaying special work by a Co-op member are held each month. Commissions accepted

MILLHAMS CRAFT STUDIO
Millhams Street
Christchurch
Dorset BH23
Contact: Mrs. D. Aldridge
Established: 1980
Open: 10am to 5pm every day
Craftworkers on site: 4
Craft: Stained glass, sculpture, pottery, painting
Prices: £1 to £200. Credit cards not accepted

COLIN BARNES
Wellow, Hambledon Road
Denmead
Hampshire PO7 6LR
Tel: 01705 255014
Contact: Colin Barnes
Established: 1994
Open: Please telephone for details
Craftworkers on site: 1
Crafts: Walking and working sticks
Prices: £40 to £200. Credit cards not accepted
Note: Upmarket pieces e.g. portraits of customer's dog etc. on a stick. Craft demonstrations arranged at county shows

BORLASE GALLERY
South Street, Blewbury
near **Didcot**
Oxfordshire OX11 9PX
Tel: 01235 850274
Contact: M.F. Ritchie
Established: 1965
Open: 10am to 12.30pm and 2.30pm to
7pm Wednesdays to Sundays. Closed
Mondays and Tuesdays
Stock: Pottery, glass, jewellery, silk
scarves, ties and waistcoats, knitwear,
greetings cards
Prices: £3 to £300. Credit cards not
accepted
Note: Exhibitions of paintings, pottery etc.
throughout year. Easy access for disabled

RUSSELL BROCKBANK
12 Highland Road
Emsworth
Hampshire PO10 7JN
Tel: 01243 431163
Contact: Russell Brockbank
Established: 1993
Open: Most days but please 'phone first
Craftworkers on site: 1
Crafts: Dumb waiters, decorative figures,
bookends, door stops
Prices: £35 to £75

SCREENS GALLERY
The Malthouse, Bridgefoot Path
Emsworth
Hampshire PO10 7EB
Tel: 01243 377334
Fax: 01243 377334
Contact: Mark Houlding
Established: 1984
Open: 10am to 1pm Tuesday, Wednes-
day, Thursday, Friday
Craftworkers on site: 2
Crafts: Screens, tables, wall panels
Prices: £250 to £10,000. Barclaycard/
MasterCard/Visa

ETON APPLIED ARTS
81 High Street
Eton
Windsor
Berkshire SL4 6AF
Tel: 01753 860771
Contact: Mike Turner
Established: 1994
Open: Monday to Saturday 10.30am to
6pm. Sunday 11am to 4pm. Closed
Wednesdays

Stock: High quality textiles, glass, studio
ceramics, domestic pottery, jewellery,
wood, metal, paper as well as wall
hangings and collage
Prices: From 50p to £1000
Note: Situated in historic Eton between
Windsor Castle and Eton College.
Gallery/workshops featuring established
makers and newcomers. Four exhibitions
annually

SUSAN WARE
24 Fairacre Rise
Fareham
Hampshire PO14 3AW
Tel: 01329 841265
Contact: Susan Ware
Established: 1983
Open: Most days but please 'phone first
Craftworkers on site: 1
Crafts: Hand painted china – from
magnets to commemorative plates and
house plaques
Prices: £2 to £50. No credit cards
Note: Disabled access

ALUM BAY GLASS LTD
The Glassworks
Alum Bay, Freshwater
Isle of Wight
PO39 0JD
Tel: 01983 753473
Fax: 01983 756262
Contact: Michael Rayner
Established: 1972
Open: 9.30am to 5pm every day except
Christmas Day
Craftworkers on site: 4
Crafts: Hand made glassware in myriad
of colours
Prices: 5p to £500. Visa/Mastercard/
Amex/Diners/JCB
Note: Glassmaking Monday to Friday all
year, famous coloured sands, chairlift to
beach in season, beautiful countryside.
Easy parking, disabled access

BRICKFIELDS HORSE COUNTRY
Newnham Road
Binstead
Isle of Wight PO33 3TH
Tel: 01983 566801/615116
Fax: 01983 562649
Contact: Mrs. Legge
Established: 1975
Open: 9am to 5pm
Craftworkers on site: 30+

Stock: Blacksmiths and farriers products.
Wooden hobby horses & rocking horses
Prices: 50p to £400. Access/Visa/
Mastercard
Note: Open all year round. Shire Horses
and miniature ponies, working displays,
wagon rides, museums and carriages.
Disabled access

QUAY ARTS CENTRE
Sea Street
Newport
Isle of Wight
PO30 5BD
Tel: 01983 528825
Contact: Ann Moorman
Established: 1984
Open: Easter to end October: Monday to
Saturday 10am to 5pm, Sunday 11am to
5pm. Winter: Monday to Saturday only
Stock: Ceramics, wood turning, mirrors
from found materials, jewellery, prints.
Note: Galleries showing visual arts
exhibitions. Disabled access ground floor
(includes shop, cafe, and workshops)

SEW-N-SO
4 Victoria Street, Ventnor
Isle of Wight
PO38 1ET
Tel: 01983 853970
Contact: Mrs. June Rostron
Established: 1979
Open: 9.30am to 12.45pm and 2pm to
5pm. Closed Mondays
Stock: Soft toys produced by owner
Speciality: Toys made to customers'
specifications (photo)
Prices: 1p to £50. Credit cards not
accepted

THEODOSIA
Holyrood Street
Newport
Isle of Wight
PO30 5AZ
Tel: 01983 525844
Contact: Veronica Chambers
Established: 1994
Open: 9.45am to 4pm. Closed Thursdays
Craftworkers on site: 2
Crafts: Silverware and jewellery in gold,
silver and gemstones
Prices: £9.50 to £2,500. Visa/Access/
Eurocard
Note: Open workshop. Commissions
accepted. Unique one-off designs

FENNY LODGE GALLERY
Simpson Road
Fenny Stratford, Bletchley
Milton Keynes
MK1 1BD
Tel: 01908 642207
Fax: 01908 647840
Contact: Mrs. S. Miller
Established: 1983
Open: 9am to 5pm Monday to Friday.
9am to 4pm Saturday. Please telephone
for details of Bank Holidays
Stock: Watercolours, oils, limited edition
prints, ceramics, designer jewellery, free
blown glass
Prices: £1.10 to £6,000. Visa/Access
Note: Selected for quality by Crafts
Council. Housed in attractive 18th
century building with beautiful garden
stretching down to Grand Union Canal

COOKIE SCOTTORN - POTTERY
Swedish House
8 Park Corner
near **Nettlebed**
Oxfordshire
RG9 6DT
Tel: 01491 641889
Contact: Cookie and John
Established: 1977
Open: Most days of week. All hours
Craftworkers on site: 2
Crafts: Countrystyle stoneware pottery,
mostly functional, plus exhibition pieces
Prices: £1 to £150. Credit cards not
accepted
Note: Located 1.5 miles from Nettlebed
along B481 on right. Advisable to phone
first but chance callers welcome.
Disabled access: gravel drive to display
room and kiln room. Other areas difficult

J.C. WOODCRAFT
54 Maple Crescent, Shaw
Newbury
Berkshire RG14 1LR
Tel: 01635 44042
Contact: John Clarke
Established: 1982
Open: Monday to Saturday 10am to 5pm
Craftworkers on site: 1
Crafts: Garden furniture and pet housing
Prices: £39 to £500. Credit cards not
accepted
Note: Member of Hants and Berks Guild
of Craftsmen - (see page 83). Individual
items made to order

CHRIS HICKS - BOOKBINDER
64 Merewood Avenue
Sandhills
Oxford
Oxfordshire OX3 8EF
Tel: 01865 69346
Contact: Chris Hicks
Established: 1981
Open: Any reasonable time but advisable
to telephone first
Craftworkers on site: 1
Crafts: Book binding and book repairs
Prices: £20 to £500. Credit cards not
accepted

OXFORD GALLERY
23 High Street
Oxford
Oxfordshire
OX1 4AH
Tel: 01865 242731
Contact: Deborah Elliott and Michelle
Bowen
Established: 1968
Open: 10am to 5pm Monday to Saturday
(occasionally closed 1pm to 1.30pm for
lunch)
Stock: Ceramics, glass, wood, contempo-
rary jewellery, textiles and limited edition
prints
Speciality: Ceramics, jewellery and prints
Prices: £7 to £4,000. Visa/MasterCard/
Amex/Diners
Note: Changing programme of ten
exhibitions a year

THE ANDEAN CRAFT CENTRE
No. 5 First Avenue
The Covered Market
Oxford
Oxfordshire OX1 3DU
Tel: 01865 790031
Fax: 01235 820007
Contact: Derek Dunn
Established: 1990
Open: 9.30am to 5.30pm Monday to
Saturday. Closed Sundays
Stock: South American clothing, silver
and costume jewellery, ceramics, wall
hangings, carpets, bags and backpacks,
musical instruments, leather furniture
Speciality: Peruvian 100% Alpaca
jumpers and cardigans
Prices: £1 to £250. Visa/Mastercard/
Amex/Switch/JCB
Note: We are importers and trade
enquiries are welcome

POOLE PRINTMAKERS
5 Bowling Green Alley
Poole
Dorset BH15 1AG
Tel: 01202 393776
Contact: John Liddell
Established: 1990
Open: Monday, Wednesday, Friday
10am to 1pm
Craftworkers on site: 12/15
Crafts: Limited edition artists' prints
Note: Practical workshop equipment
being used by members producing
original prints, screen, relief, intaglio etc
Note: Not good access for disabled

JIM CROCKATT FURNITURE
Pococks Cottage, Mariners Lane
Bradfield
near **Reading**
Berkshire RG7 6HX
Tel: 01734 744728
Contact: Jim Crockatt
Established: 1987
Open: 9am to 5pm
Craftworkers on site: 3
Crafts: Many styles of wooden furniture
from the traditional Georgian bookcase to
organic chests and 'whacky' CD holders
Prices: £50 to £3000. No credit cards
Note: Not a theme park but children are
very welcome

JOHN NIXON FINE FURNITURE
15 The Street, Aldermaston
near **Reading**
Berkshire RG7 4LN
Tel: 0118 971 3875
Fax: 0118 971 3875
Contact: John Nixon
Established: 1991
Open: 9am to 5pm by appointment
Craftworkers on site: 1/2
Crafts: Bespoke furniture
Prices: £100 to £10,000. No credit cards
Note: Member of Hampshire & Berkshire
Guild of Craftsmen & Crafts Council list

BETTLES GALLERY
80 Christchurch Road
Ringwood
Hampshire BH24 1DR
Tel: 01425 470410
Fax: 01425 479002
Contact: Gill Bettle
Established: 1989
Open: 10am to 5pm Tuesday to Friday.

10am to 1pm Saturday
Stock: Studio ceramics by British makers.
Contemporary paintings
Prices: £10 to £1000. Barclaycard etc.
accepted
Note: 8/9 solo or group exhibitions held
per year. Crafts Council listed. Slightly
restricted disabled access due to 2 steps

HERITAGE WOODCRAFT
Unit 5, Shelley Farm
Ower
near **Romsey**
Hampshire SO51 6AS
Tel: 01703 814145
Fax: 01703 814145
Contact: David Smith
Established: 1991
Open: 8.30am to 5.30pm Monday to
Friday. Weekends by appointment
Craftworkers on site: 4
Crafts: Hardwood garden furniture and
traditional wooden wheelbarrows
Prices: £28 to £1,600. Visa/Mastercard
Note: All designs can be adapted to suit
customer needs. Commissions accepted.
Situated on working farm. Workshop
open for inspection

CRAFT STUDIO
4 Gold Hill Parade
Shaftesbury
SP7 8LY
Tel: 01747 854067
Contact: Chris Morphy
Established: 1981
Open: 9.30am to 12.30pm and 1.30pm
to 5.30pm daily. Closed Wednesdays
Craftworkers on site: 2
Crafts: Silverware and jewellery in gold,
silver and gemstones
Prices: £3.50 to £1,500. Visa/Access/
Eurocard/American Express
Note: Open workshop. Commissions
welcome. Unique one-off designs a
speciality. Repairs undertaken

ALPHA HOUSE GALLERY
Alpha House, South Street
Sherborne
Dorset DT9 3LU
Tel: 01935 814944
Fax: 01935 814617
Contact: Tim and Fay Boon
Established: 1991
Open: Tuesday to Friday 10 am to
4.30pm. Saturdays 10am to 5.30pm

Crafts: High quality studio ceramics,
turned wood, glass as well as original
paintings and sculpture
Prices: From £25 to £2000. All major
credit cards accepted
Note: Regular exhibition held

MARTIN TURNER WOODWORK
'At the Sign of the Parrot'
Digby Court, Digby Road
Sherborne
Dorset DT9 3NL
Tel: 01935 813219
Fax: 01935 813219
Contact: Martin Turner
Established: 1989
Open: Please telephone for times
Craftworkers on site: 2
Crafts: Fine carved and sculpted
woodwork sometimes inlaid with semi-
precious materials
Prices: £50 to £2,500. Credit cards not
accepted
Note: Courses in wood carving and
sculpting held monthly. Talks arranged

STATION ARTISTS
Station Road
Sherborne
Dorset DT9 3NB
Tel: 01935 816618
Contact: C. McLeod, L. Belbin
Established: 1995
Open: Please telephone for times
Craftworkers on site: 2
Crafts: Ceramics and wide variety of
paintings
Prices: £5 to £2,000. Credit cards not
accepted
Note: Workshop on Sherborne Station
platform on the Waterloo to Exeter line.
Small gallery area. Disabled access via
station platform

PATRICIA STAPLES CRAFTS
73 Provene Gardens, Waltham Chase
Southampton
Hampshire SO32 2RW
Tel: 01489 896443
Contact: Patricia Staples
Established: 1996
Open: Most week days but telephone first
Craftworkers on site: 1
Crafts: Vases, candlesticks, light catchers
and jewellery made from mostly recycled
bottles and wire
Prices: £1.50 to £8. No credit cards

THE FIRST GALLERY
1 Burnham Close
Bitterne
Southampton
Hampshire SO1 5DG
Tel: 01703 462723
Established: 1984
Open: Regular hours during exhibitions
but please telephone at other times
Stock: Ceramics and fine art pictures
Note: Permanent changing exhibition

BROUGHTON CRAFTS
High Street
Stockbridge
Hampshire SO20 6HB
Tel: 01264 810513
Contact: Dick Pugh
Established: 1982
Open: 9.30am to 5.30pm Monday to
Saturday. Closed daily 1pm to 2pm
Stock: Variety of crafts
Prices: 10p to £600. Visa/Access

THEALE FIREPLACES
Mile House Farm
Bath Road
Theale
near Reading
Berkshire RG7 5HJ
Tel: 01734 302232
Fax: 01734 323344
Contact: J. Woosnam
Open: 8am to 5pm weekdays. 10.30am
to 4pm Saturdays
Craftworkers on site: 6/10
Crafts: Woodwork and marblework
Note: We manufacture to order without
surcharge to the customer

ANNE BROOKER CRAFT SHOP
Upstairs at The Lamb Arcade
High Street
Wallingford
Oxfordshire OX10 0BX
Tel: 01491 833800
Contact: Anne Brooker
Established: 1980
Open: Monday, Tuesday, Thursday and
Friday 9.30am to 5pm. Wednesday 10am
to 4pm. Saturday 9.30am to 5.30pm
Stock: Ceramics, jewellery, hand blown
glass, cards and traditional wooden toys
Prices: £1.50 to £90. Visa/Mastercard/
Amex/Diners
Note: Coffee shop & wine bar also in the
Arcade which is primarily of antique shops

BLENHEIM POTS & PLAQUES
Blenheim Farm
near Benson
Wallingford
Oxfordshire
OX10 6PR
Tel: 01491 839707
Contact: Lucienne de Mauny
Established: 1984
Open: Wednesday to Saturday 10am to
5pm
Craftworkers on site: 2
Crafts: Hand-thrown slip-decorated
tableware. Exterior plaques for names,
numbers and signs. Commemorative
ware to order
Prices: £2.50 to £2,000. Credit cards not
accepted
Note: Tranquil location. Showroom hours
as above but please telephone prior to
visiting if travelling long distances.
Visitors to workshop by appointment

SANDY HILL WORKSHOPS
Sandy Hill Lane
Corfe Castle
Wareham
Dorset BH20 5JF
Tel: 01929 480977
Fax: 01929 480977
Contact: Tony Viney
Established: 1984
Open: Variable – please telephone for
details
Craftworkers on site: 5
Crafts/stock: Polished stone plates and
bowls, stone sculpture, joinery, picture
framing, fine arts
Prices: £5 to £400. Credit cards not
accepted
Note: Adjacent to steam railway station

ANNE GOBLE SMOCKING
40 Windmill Close
Clanfield
Waterlooville
Hampshire
PO8 0NA
Tel: 01705 571425
Contact: Mrs. Anne Goble
Open: Please telephone for details
Crafts: Children's dresses etc. Household
items
Prices: £3 to £50. Credit cards not
accepted
Note: Demonstrations given of smocking
pleater as well as smocking

CLANFIELD POTTERY
131 Chalton Lane
Clanfield
Waterlooville
Hampshire
PO8 0RQ
Tel: 01705 595144
Contact: Sarah Mulley
Established: 1983
Open: 10.30am to 5pm daily. Closed
Wednesday afternoons
Craftworkers on site: 3
Crafts/stock: Pottery – specialists in
terracotta garden pots, glazed planters,
ceramic lamps and ceramic house name
plaques. Also small ceramic gifts, dried
flowers, essential oils and lampshades
Prices: £1 to £200. Visa/MasterCard
Note: Small family run pottery. Profes-
sional member of the Craft Potters
Association. Sales held most bank holiday
weekends.

DANSEL GALLERY
Rodden Row
Abbotsbury
Weymouth
Dorset
DT3 4JL
Tel: 01305 871515
Fax: 01305 871518
Contact: Danielle Holmes
Established: 1979
Open: Seven days a week. 10am to
5.30pm spring and autumn. 9.30am to
6pm summer. 11am to 4pm winter
weekdays. 10am to 5pm winter week-
ends
Stock: Woodwork only including
furniture, bowls, boxes, clocks, mirrors,
jewellery, domestic ware, carving,
children's toys and jigsaws
Speciality: Contemporary woodwork
Prices: £1 to £3,000. Access/Visa/Switch
Note: Everything is handmade by
selected UK craftsmen and is unique.
Located in old thatched converted
stableblock in very picturesque village by
the sea

EDWARD HARLE – WOODCRAFT
Selrah'
44 Micheldever Road
Whitchurch
Hampshire
RG28 7JH
Tel: 01256 892875

Contact: Edward Harle
Established: 1995
Open: Visitors welcome most days by
appointment
Crafts: High quality irregularly shaped
wooden bowls, platters, dishes, trays
and boxes carved primarily from English
hardwoods. Abstract sculptures and
small gift items. Very large bowls and
platters
Prices: £2.50 to £350. Visa/MasterCard
Note: Commissions undertaken

WHITCHURCH SILK MILL
28 Winchester Street
Whitchurch
Hampshire
RG28 7AL
Tel: 01256 893882
Contact: Mrs. C.D. Beresford
Established: 1800
Open: Tuesday to Sunday 10.30am to
5pm all year
Craftworkers on site: 6
Crafts: Silk woven on Victorian looms for
interior design, historic houses and
theatrical costume. Hand crafted gifts
Prices: 50p to £40. Switch/Amex/ Visa/
Access/MasterCard
Note: Tearooms serving light lunches;
gardens by the River Test; demonstrations
of silk weaving. Partial disabled access

HARE LANE POTTERY
Cranborne
near **Wimborne**
Dorset
BH21 5QT
Tel: 01725 517700
Contact: Jonathan Garratt
Established: 1981
Open: Usually open, including week-
ends, but please check by telephone
Craftworkers on site: 1
Crafts: Frostproof terracotta plant pots
and glazed earthenware all woodfired
and made from locally dug clay
Prices: £1 to £250. Credit cards not
accepted
Note: 18th century style pottery with
large wood fired kiln fired every seven
weeks. Attractive old dairy yard with
numerous pots planted up. Glazed ware
in old cowshed. Cranborne Garden
Centre (famous for old roses) five minutes
away. Also citrus and bamboo nurseries
locally

HONEYBROOK FORGE
High Honeybrook Farm
Cranborne Road
Wimborne
Dorset BH21 4HW
Tel: 01202 848676
Contact: Mr. C. Perry
Established: 1990
Open: 9am to 4.30pm. Saturday 9am to 12.30pm
Craftworkers on site: 2+
Crafts: Metal furniture, interior items, garden furniture and features
Note: Interesting and complex commissions are ongoing and our 18th century gallery and forge is open to the public. Close to the old town of Wimborne in an area of natural beauty

STAPEHILL ABBEY CRAFTS & GARDENS
Wimborne Road West
Stapehill
Wimborne
Dorset BH21 2EB
Tel: 01202 861686
Contact: Mrs. Della Pickard
Established: 1992
Open: 10am to 5pm Easter to October. 10am to 4pm Wednesday to Sunday November to Easter
Craftworkers on site: 20
Crafts/stock: Glass blowing, pottery, ceramics, silver jewellery, découpage, costumes, model railways, glass engraving, heraldry, dried flowers, candle making, wrought iron work
Prices: £2 to £100+. Visa in craft units
Note: Craft demonstrations held. Old Abbey, agricultural museum, beautiful gardens. Limited disabled access. Craft Fairs held with visiting crafts people

THE EDMONDSHAM POTTERY
Small Bridge Farm House
Edmondsham, Cranborne
near **Wimborne**
Dorset BH21 5RH
Tel: 01725 517251
Contact: Timothy Paul Dancey
Established: 1981
Open: 9am to 5pm
Craftworkers on site: 1
Crafts: Pottery, stoneware and terracotta
Prices: £1 to £22.50. Credit cards not accepted
Note: No charge for admission and visitors can see pots being made

WALFORD MILL CRAFT CENTRE
Stone Lane
Wimborne
Dorset BH21 1NL
Tel: 01202 841400
Contact: Maureen Barnatt and Carol Smith
Established: 1986
Open: 10am to 5pm 7 days a week except Christmas. Closed Mondays January to March
Stock: Contemporary British crafts – all media
Prices: £2 to £500. All major credit cards accepted
Note: Crafts Council selected. Licensed restaurant, gardens, river, gallery with changing exhibitions including furniture, weaving and jewellery workshops with makers in residence. Courses held

WE THREE KINGS
19 Bridge Street
Witney
Oxfordshire OX8 6DA
Tel: 01993 775399
Contact: Bill and Nikki Maddocks
Established: 1995
Open: Tuesday to Saturday 10am to 4pm. Thursday 10am to 2pm
Craftworkers on site: 1
Crafts/stock: Jewellery, ceramics, glass and wood
Prices: £5 to £5,000. Visa/Mastercard
Note: Working goldsmith on site. Regular exhibitions throughout the year. Gallery stocks work from new and established designer craftsmen. Located in charming 17th century listed building

THE CRAFT SHOP
Ashridge Manor Farm
Forest Road
Wokingham
Berkshire RG11 5QY
Tel: 01734 770816
Contact: Beryl Chapman
Established: 1990
Open: 10am to 4.30pm. Closed Mondays
Stock: Hand made crafts, ceramics, pictures, silk and dried flowers, jewellery, cards, bric-à-brac, candles, gifts
Speciality: Most crafts made to customers' own specifications
Prices: £1 to £85. Credit cards not accepted
Note: Creative woodworking shop nearby

CHAPEL YARD POTTERY
West Street
Abbotsbury
Dorset
DT3 4JR
Tel: 01305 871663
Contact: Richard Wilson
Established: 1990
Open: 9.30am to 5.30pm Tuesday to Saturday
Craftworkers on site: 1
Crafts: Domesticware pottery from mugs to casseroles
Prices: £2.50 to £80. Credit cards not accepted
Note: Customers can see pots made on the premises. Plenty of parking available. Abbotsbury Village has many other attractions for the visitor

BEAUX ARTS
12/13 York Street
Bath
BA1 1NG
Tel: 01225 464850
Fax: 01225 422256
Contact: Colin Carlin
Established: 1979
Open: 10am to 5pm Monday to Saturday
Stock: Very fine studio pots and modern ceramics. Collectors pieces
Speciality: Contemporary ceramics
Prices: £50 to £3,000. Mastercard/Visa/American Express
Note: Situated in the heart of Bath close to Roman baths

HITCHCOCK'S
10 Chapel Row
Bath
BA1 1HN
Tel: 01225 330646
Fax: 01225 330646
Contact: Fleur Hitchcock
Established: 1986
Open: 10am to 5.30pm Monday to Saturday
Stock: Contemporary ceramics, jewellery, textiles, automata
Speciality: Knitwear and wooden mechanical toys
Prices: £10 to £500. Access/Visa/Switch
Note: The gallery has an emphasis on new makers and holds regular themed exhibitions

PETER HAYES CONTEMPORARY CERAMICS
2 Cleveland Bridge
Bath
BA1 5DH
Tel: 01225 466215
Fax: 01225 311233
Contact: Peter Hayes
Established: 1982
Open: 10am to 5pm
Craftworkers on site: 4
Crafts: Individually made ceramics and sculpture
Prices: £100 to £3,000. Access/Barclaycard
Note: Sculpture garden and raku demonstrations

ST. JAMES'S GALLERY
9B Margaret's Buildings
(near the Royal Crescent)
Bath BA1 2LP
Tel: 01225 319197
Contact: R. Sloman
Established: 1983
Open: 10am to 5.30pm Monday to Saturday
Stock: Modern ceramics, hand made jewellery and paintings all by British artists
Prices: £5 to £2000. Mastercard/Visa/American Express
Note: Crafts Council recommended gallery. Ground floor access to disabled

EELES FAMILY POTTERY
Shepherds Well
(A3066) at Mosterton
near **Beaminster**
Dorset
DT8 3HN
Tel: 01308 868257
Contact: David Eeles
Established: 1961
Open: 9am to 6pm daily
Craftworkers on site: 5
Crafts: Wood fired ceramics, stoneware and porcelain
Prices: £2 to £300. Credit cards not accepted
Note: Open days at pottery on weekends in the spring and autumn for demonstrations, conducted tours of the workshops and wood fired dragon kiln. See entry for Eeles Pot Shop in Lyme Regis

RONALD EMETT: FURNITURE MAKER
2 The Square
Beaminster
Dorset DT8 3AS
Tel: 01308 863000
Contact: Ronald Emett
Established: 1990
Open: Workshop not generally open.
Gallery open 10.30am to 5.30pm
Monday to Saturday during holiday
season. Otherwise 10.30am to 5.30pm
Friday and Saturday
Craftworkers on site: 1
Crafts: Furniture
Prices: £500 to £5,000. Visa/Delta/
Access/Mastercard
Note: Gallery stocks ceramics, prints,
paintings, silver, silk from £5 to £300.
Workshop open days on bank holiday
Mondays. Disabled access easy

THE CRAFT & DESIGN CENTRE
Broadwindsor
Beaminster
Dorset DT8 3PX
Tel: 01308 868362
Contact: Kate Guilor
Established: 1986
Open: 10am to 5pm everyday
Craftworkers on site: 9
Crafts: Hats, knitwear, weather signs,
wood turning, joinery, upholstery,
watercolours
Prices: 50p to £100+. Credit cards not
accepted
Note: A range of shops, workshops and
studios situated in beautiful countryside.
Spacious restaurant. Disabled access.
Local attractions include Parnham House,
Forde Abbey and Horn Park

APPLEDORE CRAFTS COMPANY
5 Bude Street
Appledore
Bideford
Devon EX39 1PS
Tel: 01237 423547
Contact: Penny Lucraft
Established: 1991
Open: Winter: 10am to 4pm Wednesday
to Sunday. Summer: 10am to 6pm seven
days
Stock: Furniture, ceramics, textiles,
bookbinding, sculpture and cards by
North Devon's leading contemporary
craftsmen
Speciality: Fine furniture, wood crafts

Prices: £3 to £5,000. Visa/Mastercard etc.
Note: Programme throughout year of
guest artist exhibitions. Direct commissioning service. Pubs and teashops in
village

DAVID CHARLESWORTH FURNITURE MAKER
Harton Manor, Hartland
Bideford
Devon EX39 6BL
Tel: 01237 441288
Contact: D. Charlesworth
Established: 1976
Open: 9am to 12 noon
Crafts: One-off commissions of furniture
Prices: £400 to £10,000. Credit cards not
accepted
Note: Full-time furniture making courses
for adults of all ages. Flexible start and
duration

MILLTHORNE CHAIRS
10 Fore Street, Hartland
near **Bideford**
North Devon
EX39 6BD
Tel: 01237 441590
Contact: Bob and Sue Seymour
Established: 1979
Open: 8.30am to 6pm Monday to
Saturday
Craftworkers on site: 3
Crafts: Children's and full size Windsor
chairs and stools hand made from elm,
ash and beech on the premises. Framed
abstract beach photos
Prices: £9 to £50. Credit cards not
accepted

SPRINGFIELD POTTERY
Springfield, Hartland
near **Bideford**
Devon EX39 6BG
Tel: 01237 441506
Contact: Philip and Frannie Leach
Established: 1979
Open: Monday to Saturday 9am to 6pm
Craftworkers on site: 3
Crafts: Decorative earthenware pottery,
oven and tableware, some tiles and
garden pots
Prices: £3 to £70. Access/Visa/Delta/
Mastercard
Note: Demonstrations are not given;
however there are other craft workshops
in the village.

WENFORD BRIDGE POTTERY
St. Breward
Bodmin
Cornwall PL30 3PN
Tel: 01208 851038
Contact: Mr. A. Cardew
Established: 1939
Open: 9am to 5.30pm
Craftworkers on site: 3
Crafts: Hand made wood fired stoneware
and ceramics
Prices: £5 to £500. Visa/Mastercard
Note: This is a particularly important
pottery in history of craft pottery in UK;
founded by Michael Cardew. Tearoom,
gallery and museum nearby

THE OLD MILL
Boscastle Harbour
Boscastle
Cornwall PL35 0HQ
Tel: 01840 250230
Contact: Barry Knowles
Established: 1987
Open: Winter: 10.30am to 5.30pm.
Summer: 9.30am to 7.30pm
Stock: Dried flowers, mohair, collectors
bears, paintings and prints, old and new
books, antiques and linens, bridal gowns
and designer fashions
Speciality: Resident dried flower and bear
artists and fashion/bridal designer
Prices: 50p to £5,000. Visa/Access/Amex
Note: Millers Pantry open daily for cream
teas and after dark the Bistro offers
sumptuous home made fayre

THE DEVON GUILD OF CRAFTSMEN
Riverside Mill
Bovey Tracey
Devon TQ13 9AF
Tel: 01626 832223
Contact: Annette Vaitkus
Established: 1986
Open: 10am to 5.30pm daily (closed
winter bank holidays only)
Stock: Textiles, ceramics, furniture, glass,
wood turning and carving, jewellery,
millinery, prints, quilts, toys
Prices: £5 to £1,000. Access/Visa/Switch
etc.
Note: Seven thematic exhibitions are held
throughout the year. Egon Ronay
recommended granary cafe serving home
cooked meals, cream teas and local
wines. Free car parking. Picturesque
setting. Crafts Council listed

SOMERSET LEVELS BASKET & CRAFT CENTRE LTD
Lyng Road
Burrowbridge
near **Bridgwater**
Somerset TA7 0SG
Tel: 01823 698688
Contact: Ruth Loveridge
Established: 1864
Open: 9am to 5.30pm Monday to
Saturday and 12.30pm to 5.30pm
Sundays all year
Craftworkers on site: 6
Crafts: Basketware of all types, cane
furniture, corn dollies, wooden items,
'silk' flowers, straw hats etc.
Prices: 20p to £1,000. Access/Mastercard/
Eurocard/Visa/Delta
Note: Craft demonstrations, display of
pictures and other items depicting King
Alfred's life, tourist information, tearoom.
Disabled access

A.R. DESIGNS
66 Providence Lane, Long Ashton
Bristol
BS18 9DN
Tel: 01275 392440
Contact: Alan Ross
Established: 1975
Open: 9am to 1pm weekdays
Craftworkers on site: 1
Crafts: Furniture to order. Range of stools
in stock
Prices: £25 to £90. Credit cards not
accepted
Note: Lovely garden

CHURCH HOUSE DESIGNS
Broad Street, Congresbury
near **Bristol**
BS19 5DG
Tel: 01934 833660
Contact: Lorraine Coles
Established: 1986
Craftworkers on site: 1
Open: Monday to Saturday 10am to 1pm,
2.15pm to 5.30pm. Closed Wednesdays
and Sundays
Crafts/stock: High quality studio
ceramics, domestic pottery, glass, wood,
furniture, metal, jewellery, textiles,
knitwear, toys as well as original prints
Prices: Under £5 to £700. Visa/Access
Note: Ceramics and glassware exhibition
twice yearly. Furniture is made here to
commission only

COUNTRY CRAFTS
55 Broad Street
Chipping Sodbury
Bristol
BS17 6AD
Tel: 01454 327157
Contact: Jenni Bull
Open: 10am to 6pm Monday to Friday.
9.30am to 5.30pm Saturday. 2pm to 5pm
Sunday
Stock: Large range of crafts. Jewellery
findings, some craft materials, i.e. glass
and ceramic paints, and craft kits
Prices: 5p to £1,000. American Express/
Cardnet
Note: Art exhibitions and craft demon-
strations when advertised

ROY YOUDALE
6 Highbury Road
Horfield
Bristol
BS7 0BZ
Tel: 0117 951 1421
Contact: Roy Youdale
Established: 1994
Open: Most days but please 'phone first
Craftworkers on site: 1
Crafts: Domestic willow baskets
Prices: £13 to £90. Credit cards not
accepted

BROADWINDSOR CHAIRS & CRAFTS
Old Exchange Workshop
Drimpton Road
Broadwindsor
Dorset
DT8 3RS
Tel: 01308 868909
Contact: Graham Wilkinson
Established: 1983
Open: 9.15am to 1pm and 2pm to
5.30pm Monday to Friday
Craftworkers on site: 1
Crafts: Caning, rushing, seagrass,
children's chairs, stools
Note: Caning or rushing may often be
seen by visitors. Visiting out of hours by
appointment

MORWENSTOW POTTERY
Lower Stursdon
Morwenstow
Bude
Cornwall
EX23 9HU
Tel: 01288 331412

Contact: Marylyn Hyde
Established: 1990
Open: During the season, otherwise by
arrangement.
Craftworkers on site: 1
Crafts: Mainly hand built stoneware pots.
Some thrown earthenware.
Prices: £5 to £350. Credit cards not
accepted

OTTERTON MILL CENTRE
near **BUDLEIGH SALTERTON**

Devon
EX9 7HG
Tel: 01395 568521
Contact: Desna Greenhow
Established: 1977
**Open: 10.30am to 5.30pm daily mid-
March to November. 11am to 4pm
daily November to mid-March**
**Craftworkers on site: 3 resident and 9
in co-operative shop**
**Crafts: Turned wood, pottery, hand-
printed clothes, furniture, toys, prints,
pressed flower pictures and mats,
jewellery, ironwork, sculpture**
**Prices: £2 to £1000. Visa/Access/
Mastercard**
**Note: At least 4 craft exhibitions are
held each year. Demonstrations and
workshops held in connection with
exhibitions. Restaurant, bakery,
working water mill producing
stoneground flour. Disabled access**

CARNON DOWNS POTTERY
Carnon Downs Garden Centre
Quenchwell Road
Carnon Downs
near Truro
Cornwall
TR3 6LN
Tel: 01872 863374
Contact: Hugh West/Barry Hugget
Established: 1986
Open: 10am to 5pm 7 days a week
Craftworkers on site: 2
Crafts: Salt glaze and Raku pottery
Prices: 50p to £1,000. Access/Visa
Note: Located in garden centre workshop
open to public. Potters can be seen at
work. Garden centre has restaurant,
children's play area, pet shop, retail
outlets including garden machinery.

CHITTLEHAMPTON POTTERY
Chittlehampton
North Devon EX37 9PX
Tel: 01769 540420
Contact: Roger and Ros Cockram
Established: 1986
Open: 10am to 1pm and 2pm to 5.30pm weekdays. Some weekends but telephone for details
Craftworkers on site: 2/3
Crafts: Handmade ceramics and pottery. Some individual pieces. Domestic stoneware, candlesticks and figures
Prices: £3 to £700. Visa/Mastercard/ Access/Eurocard
Note: Visitors may tour studios next to showroom. Parking available. The pottery is on the edge of a beautiful ancient village with wonderful church and pub in Devon countryside

CLEVEDON CRAFT CENTRE
Moor Lane
Clevedon
North Somerset
BS21 6TD
Tel: 01275 342114
Contact: Mr. D.A. Stear
Established: 1971
Open: Tuesday to Sunday all year
Craftworkers on site: 10
Crafts: Silver and gold jewellery, silverware, glass engraving, pottery, painting, sign writing, arashi shitsori, silk flowers, toys, knitting, dried flowers, leatherwork, craft jewellery
Prices: £2 to £2,000. Amex/Access/Visa accepted by some workshops
Note: Tearoom, gallery, geese, ducks, hens. Free entry and parking. Disabled access to some workshops

ART BENATTAR CRAFT
31 Market Square
Crewkerne
Somerset
TA18 7LP
Tel: 01460 73118
Contact: Elsa Benattar
Established: 1994
Open: 10am to 5pm Monday to Saturday
Craftworkers on site: 1
Crafts/stock: Ceramics, wood/iron work, jewellery, hand knits, prints
Prices: £4 to £600. Credit cards not accepted
Note: Coffee shop

DARTINGTON CIDER PRESS CRAFT CENTRE
Shinners Bridge
Dartington
Totnes, South Devon TQ9 6TQ
Tel: 01803 864171
Fax: 01803 866094
Contact: Christine Graves
Established: 1997
Craftworkers on site: 1
Open: Monday to Saturday 9.30am to 5.30pm and Sundays from 10.30am to 5.30pm from Easter to Christmas
Crafts/stock: Domestic pottery, studio ceramics, glass, wood, metal, jewellery, and textiles
Prices: From £10 to £300. Most major cards accepted plus Switch
Note: Fine British crafts with strong West Country representation. Gift shops also

FACETS
Broadstone
Dartmouth
Devon TQ6 9WR
Tel: 01803 833534
Fax: 01803 833534
Contact: J.S. Dilley
Established: 1985
Open: 10am to 5pm Monday to Saturday. 10am to 1pm Wednesday during winter
Stock: High quality jewellery both precious and non-precious. Design and commission service available
Prices: £7 to £900. All major credit cards accepted
Note: Crafts Council listed

HIGHER STREET GALLERY
1 Higher Street
Dartmouth
Devon TQ6 9RB
Tel: 01803 833157
Contact: Mark Goodwin
Established: 1982
Open: 9.30am to 5pm daily. Closed Sundays January to March
Stock: High quality ceramics, sculpture, metalwork, studio glass, toys, wood, jewellery, etchings, paintings
Prices: £2.50 to £500. Most major cards plus Switch
Note: 16th Century Grade II listed building – showroom of special ambience – selected for quality by Crafts Council. Owner run gallery with accent on service and knowledge of art and crafts

SIMON DREW GALLERY
13 Foss Street
Dartmouth
Devon
TQ6 9DR
Tel: 01803 832832
Fax: 01803 833040
Contact: Caroline and Simon Drew
Established: 1981
Open: Monday to Saturday 9am to 5pm.
Open some Sundays in summer and
closed part of January – please phone to
check
Stock: Domestic pottery, studio ceramics.
Stoneware, porcelain and earthenware.
Functional, decorative and sculptural pots
as well as prints, cards and books
Prices: From £6 up to £500. Most major
credit cards and Switch accepted

WALTER'S CONTEMPORARY CRAFTS
8 Market Street
Dartmouth
Devon
TQ6 9QE
Tel: 01803 833318
Contact: Mrs. Janet Gent
Established: 1992
Open: 10am to 5pm Monday to Saturday,
11.30am to 4pm Sundays during summer
season. Shorter hours in winter (closed
Wednesdays)
Stock: Pottery, ceramics, wood, metal,
papier mâché, jewellery, textiles,
paintings, prints
Prices: £10 to £250

THE VIVIAN GALLERY
2 Queen Street
Dawlish
Devon
EX7 9HB
Tel: 01626 867254
Contact: Hazel and Peter Bunyan
Established: 1988
Open: 9am to 6pm Monday to Saturday.
10am to 3pm most Bank Holidays
Stock: "Wearable art", clothing, jewellery
and accessories, cards
Speciality: Sterling silver and bead
jewellery by Hazel Bunyan
Prices: £2 to £150. Visa/Mastercard/
Delta/Switch
Note: Occasional demonstrations and
exhibitions of crafts and makers' work.
Teashop and restaurant nearby. Restricted
access for disabled due to steps

COOMBE FARM GALLERY
Dittisham
near Dartmouth
Devon TQ6 0JA
Tel: 01803 722352
Fax: 01803 722275
Contact: Mark Riley
Established: 1992
Open: 10am to 5pm Monday to Saturday.
Sunday by appointment
Stock: Regular theme exhibitions and
displays of work in many media by some
of the best artists and makers, including
West Country artists and new designers
Prices: £5 to £2,000. Visa/MasterCard
Note: Gallery is part of Coombe Farm
Studios, a teaching centre for fine arts and
crafts. Free parking available. Poor
disabled access

ABBEY POTTERY
Cerne Abbas
Dorchester
Dorset DT2 7JQ
Tel: 01300 341865
Contact: Paul Green
Established: 1986
Open: 10am to 6pm daily (closed
Mondays in winter)
Craftworkers on site: 1
Crafts: Hand-thrown stoneware and
porcelain
Prices: £3.50 to £100+
Note: Pottery made on premises. Situated
close to the Cerne Giant, a famous chalk
figure cut into the hillside and adjacent to
the remains of Cerne Abbey, established
987 AD. Member of the Craft Potters
Association

WARWICK PARKER
The Dairy House, Maiden Newton
Dorchester
Dorset DT2 0AE
Tel: 01300 320414
Contact: Warwick Parker
Established: 1967
Open: 9am to 5pm Monday to Friday
(please telephone if coming from long
distance)
Craftworkers on site: 1
Crafts: Ceramics & painted wooden boxes
Prices: £4 to £100. Credit cards not
accepted
Note: Warwick has exhibited throughout
Britain, as well as in Germany and New
Zealand

THE STUDIO POTTERY GALLERY
15 Magdalen Road
Exeter
Devon EX2 4TA
Tel: 01392 430082
Contact: Paul Vincent
Established: 1989
Open: Saturday to Monday 11.30am to 9pm but please phone to check
Stock: Domestic pottery, studio ceramics
Prices: From £5 to £500. Most major cards accepted
Note: Monthly exhibition programme

TIM ANDREWS POTTERY
Greenway, Woodbury
near **Exeter**
Devon EX5 1LW
Tel: 01395 233475
Contact: Tim and Sandra Andrews
Established: 1993
Open: 10am to 6pm Monday to Saturday
Craftworkers on site: 1
Crafts: Individual ceramics specialising in raku and smoke fired pots. Garden pots
Prices: £4 to £400. Credit cards not accepted
Note: Annual autumn ceramics exhibition held September/October featuring leading British makers. Fellow of Craft Potters Association.

BESIDE THE WAVE GALLERY
10 Arwenack Street
Falmouth
Cornwall TR11 3JA
Tel: 01326 211132
Fax: 01326 212212
Contact: Lesley Dyer
Established: 1989
Open: Monday to Saturday 9am to 5.30pm and by appointment
Stock: Ceramics, including porcelain, earthenware and stoneware. Contemporary Cornish paintings and prints, artists' cards and limited edition prints, including etchings, silkscreens and lithographs
Note: Four or five exhibitions held each year. Located in pretty harbour town

OLD HOUSE OF FOYE
31-35 Fore Street
Fowey
Cornwall PL23 1AH
Tel: 01726 833712
Contact: Cyril and Rosemary Blackstock
Established: 1987

Open: 10am to 1pm, 2.15 to 5pm. Half day Wednesdays in winter
Stock: Pottery, jewellery, third world crafts, tea and coffee with containers such as coffe/tea pots and grinders
Speciality: Rugs and Christian books etc
Prices: 5p to £500. Access/Visa
Note: Medieval house with enchanting old kitchen which is little altered. Unspoilt ancient town surrounded by National Trust land. Great restaurants and coffee shops

THE BLACK SWAN GUILD
2 Bridge Street
Frome
Somerset BA11 1BB
Tel: 01373 473980
Contact: Ann O'Dwyer
Established: 1986
Craftworkers on site: about 7
Open: 10am to 5pm Monday to Saturday. Some Bank Holidays – please 'phone first
Crafts: Contemporary art and crafts including jewellery, basket weaving, picture framing, textiles, ceramics, glass, wood, metal, toys and original prints
Most major credit cards accepted
Prices: From £10 to over £1000 – most major cards accepted
Note: Shop and gallery with full programme of major shows. Wholefood cafe and courtyard. Disabled access on ground floor. Housed in historic building including renovated wool drying tower. The Guild has charitable status and provides workspace for young makers, with courses & educational programmes for the public

ART OF AFRICA
20 Magdalene Street
Glastonbury
Somerset BA6 9EH
Tel: 01458 835343
Fax: 01458 835343
Contact: Steve Hill and Roz Edwards
Established: 1989
Open: 10am to 6pm, 7 days a week
Stock: Finest quality art, handicrafts & musical instruments of genuine African origin
Prices: 10p to £500. Access/Visa/Mastercard/Eurocard/AmExpress
Note: The owner reports "We are rated as one of the most original and popular shops in our area". Set in the interesting & historic town of Glastonbury

HATHERLEIGH POTTERY
20 Market Street
Hatherleigh
Devon EX20 3JP
Tel: 01837 810624
Fax: 01837 810624
Contact: Elizabeth Aylmer
Established: 1984
Open: 10am to 6pm Monday to Saturday.
Sundays by appointment
Craftworkers on site: 2 full-time, 4 part-
time
Crafts: Hand thrown stoneware, domestic
and garden pots produced on premises
Prices: £1 to £100. Access/Visa/Eurocard/
Mastercard
Note: Inviting atmospheric environment
where work can be seen in progress

STOCK IN TRADE
41 Market Street
Hatherleigh
Devon EX20 3JP
Tel: 01837 810624
Fax: 01837 810624
Contact: Jenny Stock
Established: 1995
Open: 10am to 6pm Monday to Saturday.
Sunday by appointment
Craftworkers on site: 2
Crafts: Block printed fabric, crafts from
Zimbabwe and pottery
Prices: £5 to £200. Access/Visa/Eurocard/
Mastercard
Note: Fabric printing can be seen in
progress

NORMAN STUART CLARKE GLASS
The Glass Gallery
St. Erth
Hayle
Cornwall TR27 6HT
Tel: 01736 756577
Contact: Norman Stuart Clarke
Established: 1984
Open: Monday to Friday 10am to 1pm,
2.30pm to 5pm. Saturday 10am to 1pm.
Closed Sundays
Craftworkers on site: 2
Crafts: Artist made coloured glass, vases,
bowls, flasks, paperweights, perfume
bottles etc.
Prices: £20 to £1000. Visa/Delta/Electron/
Mastercard/Switch/Eurocard
Note: Visitors may see artist at work at
the furnace (glassblowing). Please make
appointment if travelling long distance

KENNACK POTTERY
Kennack Sands
near **Helston**
Cornwall
TR12 7LX
Tel: 01326 290592
Contact: Michael Hatfield
Established: 1982
Open: 9am to 5pm most days
Craftworkers on site: 2
Crafts: Ceramic models in miniature and
detail: boots, wellies, mice, etc.
Prices: 75p to £100. Credit cards not
accepted
Note: Large 'hands-on' area, craft
demonstrations, tearoom

NORMAN UNDERHILL FIGURINES
Collectors Corner
Meaver Road, Mullion
Helston
Cornwall
TR12 7DN
Tel: 01326 240667
Contact: Norman Underhill
Established: 1970
Open: 10am to 10pm summer. 10am to
6pm winter
Stock: Handmade figurines
Prices: £10 to £160. Visa
Note: Commissions taken. Situated in
pretty Cornish village

DAVID CLEVERLY POTTERY
Haytown
Holsworthy
Devon EX22 7UW
Tel: 01409 261476
Fax: 01409 261476
Contact: Margaret Cleverly
Established: 1979
Open: Easter to end of September. For
other times please telephone
Craftworkers on site: 2
Crafts: Colourful decorative and
functional pottery. Embroidery by
Margaret Cleverly
Prices: £3.50 to £300. Visa/MasterCard/
Eurocheque

HAYTOWN POTTERY
Haytown
Holsworthy
Devon EX22 7UW
Tel: 01409 261476
Fax: 01409 261476
Established: 1979

Open: Regular hours in season. Closed
Saturdays. Other times please telephone
first
Craftworkers on site: 1
Crafts: Colourful domestic pottery and
animals
Prices: £3.50 to £350. Visa/MasterCard

JOHN & JAN MULLIN
September House
Parnacott
Holsworthy
Devon
EX22 7JD
Tel: 01409 253589
Contact: John Mullin
Established: 1980
Open: 9am to 6pm Monday to Saturday
Craftworkers on site: 2
Crafts: Ceramic sculpture plus personal-
ised commissions
Prices: £12 to £750. Visa/Eurocard/
Mastercard/Access
Note: C.P.A. and Devon Guild member

FUCHSIA CERAMICS
Cutlands Pottery
Chillington
near **Kingsbridge**
South Devon
TQ7 2HS
Tel: 01548 580390
Contact: Paul and Valerie Metcalfe
Established: 1978
Open: 9am to 1pm and 2pm to 5.45pm.
High season only Saturday 9am to
12.30pm
Craftworkers on site: 1
Crafts: Hand thrown stoneware pottery
decorated with our unique fuchsia design.
Variety of crafts and prints
Prices: £1.50 to £30. Credit cards not
accepted
Note: Individual stoneware house
plaques made to order. Working pottery
where throwing, turning, kiln packing,
glazing and decorating are all carried out.
Disabled access

MUCHELNEY POTTERY
Muchelney
Langport
Somerset TA10 0DW
Tel: 01458 250324
Contact: John Leach
Established: 1965
Open: Monday to Friday 9am to 1pm,

2pm to 5pm. Saturday 9am to 1pm
Craftworkers on site: 3
Crafts: Handthrown wood fired stoneware. Oven to table designed individual
pieces
Prices: £5.30 to £300. Mastercard
Note: Pond and wildlife conservation
area. National Trust and English Heritage
properties in village

PIPPA BERTHON SILVERSMITH
Silver Workshop
Bow Street
Langport
Somerset TA10 9PQ
Tel: 01458 251122
Contact: Pippa Berthon
Open: 9.30am to 5.30pm Monday to
Friday. 9.30am to 1pm Saturday
Craftworkers on site: 2
Crafts: Small silver articles
Prices: £9 to £400
Note: Workshop is in shop so articles can
be seen in the making

EELES FAMILY POT SHOP
56 Broad Street
Lyme Regis
Dorset
Tel: 01308 868257
Contact: David Eeles
Established: 1983
Open: 9am to 5.30pm Monday to
Saturday
Crafts: Wide range of wood fired
ceramics
Prices: £2 to £300. Credit cards not
accepted
Note: The workshop for this shop is
located in Beaminster – see first page of
this section

WOODSIDE CRAFT CENTRE
Watersmeet Road
Lynmouth
North Devon EX35 6EP
Tel & Fax: 01598 752529
Contact: Mr. and Mrs. Jost
Established: 1977
Open: 10.30am to 5pm Feb. half term to
November
Stock: Local crafts and craft kits
Prices: £1.50 to £20. Access/Visa/
Mastercard/Switch
Note: Brass rubbing – over 100 facsimi-
les. Special deals for schools. Disabled
welcome. Easy parking

COUNTRY CRAFTS
16 The Square
Aldbourne
near **Marlborough**
Wiltshire SN8 2DU
Tel: 01672 40478
Contact: Richard J. Hale
Established: 1989
Open: Tuesday, Wednesday, Thursday,
Saturday 10am to 4.30pm. Friday 2pm to
4.30pm
Stock: Watercolours, minerals, dried
flower arrangements, silk flowers, glass,
greetings cards, pottery, jewellery, toys
and frames
Prices: £2 to £80. Credit cards not
accepted
Note: Situated in a village set in the
Marlborough Downs. Teashop adjoining

MICHAEL BURTON
Osborne Cottage
Hurst
Martock
Somerset
TA12 6JU
Tel: 01935 822362
Fax: 01935 822352
Contact: Michael Burton
Established: 1973
Open: 9am to 6pm
Craftworkers on site: 1-2
Crafts: Smithing – bowls etc. Carved
jewels
Prices: £2.50 to £20,000. Credit cards not
accepted
Note: Owner can be seen at work

THE SOMERSET GUILD OF CRAFTSMEN GALLERY
Yandles of Martock
Martock
Somerset
Tel: 01278 424983
Contact: Ron Rudd
Established: 1994
Open: Monday to Saturday 9am to 5pm.
Sundays 11am to 5pm
Stock: Musical string instruments,
spinning wheels, wood turning and
furniture, basket work, ceramics, textiles,
silver, metalwork, carved wooden birds
Prices: From £10 to £700. Credit cards
accepted
Note: This gallery is staffed each day by
a Guild Member. Demonstrations held on
nominated days throughout the year

D.B. POTTERY
Highway Cottage
Church Street
Merriott
Somerset
TA16 5PR
Tel: 01460 75655
Contact: David Brown
Established: 1986
Open: Weekends and daily throughout
July and August. Other times vary –
please telephone
Craftworkers on site: 1
Crafts: Hand built and thrown pottery
Prices: £2 to £200. Credit cards not
accepted
Note: Demonstrations on request.
Disabled access

MILL POTTERY
Wootton Courtenay
Minehead
Somerset
TA24 8RB
Tel: 01643 841297
Contact: Michael Gaitskell
Established: 1974
Open: 9.30am to 1pm and 2pm to
5.30pm
Craftworkers on site: 1
Crafts: Top quality reduction fired
stoneware, studio individual pots and
wide choice of everyday items
Prices: £5 to £250. Credit cards accepted
Note: Very well established, many repeat
visitors and recommendations, distinctive
personal but classical style. Fully
accessible workshop including demon-
strations. Interesting kiln. Fully operative
hydro-electric and mechanical
waterwheel. Easy car parking

PARK STREET POTTERY & GALLERY
13A Park Street
Minehead
Somerset
TA24 5NQ
Tel: 01643 702753
Contact: Maureen Shearlaw
Established: 1992
Open: Monday to Friday 10am to 1pm
and 2pm to 5pm. Saturday 10am to 1pm
Craftworkers on site: 1
Crafts/Stock: Unique one-off porcelain,
mostly highlighted in pure gold. Also
functional items in stoneware. High
quality hand blown glass, sculptural

woodturning, bronze sculptures etc.
Prices: £3 to £300. All major credit cards accepted
Note: Pottery can be seen being produced at all times. Member of Devon Guild of Craftsmen. Founder member of West Country Potters Association

WOODTURNERS (SOUTH DEVON) CRAFT CENTRE
New Road
Modbury
South Devon PL21 0QH
Tel: 01548 830405
Contact: Mr. J. Trippas and Mrs. I. Trippas
Established: 1959
Open: 10.30am to 5.30pm daily except Sundays. By appointment only from 23rd Dec. to 12th Feb.
Craftworkers on site: 2
Crafts: Furniture, tables, chairs, dressers, units – all in hardwoods. Fruit and salad bowls, all types of woodturning, wall carvings, house names. Wooden toys, rocking horses, pottery, baskets, sheep-skins, slippers, silver, pressed flower pictures – locally made
Prices: 20p to £6,000+. Credit cards not accepted
Note: Commissions taken. Visitors may see furniture being made or woodturning and possibly carver at work. Parking available. Centre located on the A379 Kingsbridge Road (leaving Modbury look for red cartwheels). Limited disabled access as upper floors not accessible to wheelchair users

ABBOT POTTERY
Hopkins Lane
Newton Abbot
Devon TQ12 2EL
Tel: 01626 334933
Contact: Caroline and Richard Smith
Established: 1987
Open: 10am to 5pm Monday to Saturday. If making special journey Saturday pm, please ring first
Craftworkers on site: 3
Crafts: Handthrown slip decorated red earthenware mugs, jugs, platters, vases
Prices: £2.50 to £50. Mastercard/Visa
Note: Pottery decorated with rural scenes complete with sheep. Other designs also on sale. Visitors can see throwing, turning, decorating. Commissions undertaken

DEVON GUILD OF CRAFTSMEN
Riverside Mill
Bovey Tracey
Newton Abbot
Devon
TQ13 9AF
Tel: 01626 832 223
Contact: Annette Vaitkus
Established: 1986
Open: Daily including weekends 10am to 5.30pm
Crafts/stock: Domestic pottery, studio ceramics, glass, wood, metal, jewellery, textiles, knitwear, toys and furniture
Prices: From £7 to £7000. Most major credit cards
Note: Wide range of craftwork from the south west region and one-person and thematic exhibitions

OTTERTON MILL POTTERY
Otterton Mill
Otterton
Devon
EX9 7HJ
Tel: 01395 567041
Contact: Keith Smith
Established: 1980
Open: 10.30am to 5.30pm April to September
Craftworkers on site: 1
Crafts: Ash glazed stoneware, mostly food orientated. Shrubs, alpines, herbaceous plants, pine shelves
Prices: £1 to £100. Credit cards not accepted
Note: Crafts Council listed. Restaurant, other workshops, shop (co-operative), working water mill and courtyard

MID-CORNWALL GALLERIES
St. Blazey Gate
Par
Cornwall PL24 2EG
Tel: 01726 812131
Contact: Helen Gould
Established: 1980
Open: Monday to Saturday 10am to 5pm
Crafts/stock: High quality domestic pottery, studio ceramics, wood, metal, jewellery, glass, silk as well as limited edition prints and paintings
Prices: From under £5 to over £400. Visa/MasterCard accepted
Note: 8/9 exhibitions held a year. Small coffee shop. Wheelchair access. Crafts Council selected gallery

GREEB COTTAGE
Greeb Cottage, Lands End
Sennen
Penzance
Cornwall, TR19 7AA
Tel: 01736 871501
Contact: Tom Brooke
Established: 1983
Open: 10.30am to 5.30pm daily
Craftworkers on site: 5
Crafts: Engraved glass, handworked
leather, jewellery, bespoke carved names
and house signs in wood
Prices: £1 to £250. Credit cards not
accepted
Note: Small animal farm for children,
marvellous cliffside location. Disabled
access. Situated 300m south of Land's
End Visitors Complex – easy walking

BRAYBROOKE POTTERY
17 Andover Road, Upavon
Pewsey
Wiltshire SN9 6AB
Tel: 01980 630466
Contact: Sally Lewis
Established: 1972
Open: Most times but please 'phone first
Craftworkers on site: 1
Crafts: Stoneware and earthenware
thrown pots. Domestic, decorative and
commemorative pots, ranging from egg-
cups through jugs, teapots, wedding
platters to fountains and fire-surrounds.
Commissions undertaken
Prices: £2.50 to £300. Cheques accepted
with bank card

COMPTON MARBLING
Lower Lawn Barns
Tisbury
Salisbury
Wiltshire SP3 6SG
Tel: 01747 871147
Fax: 01747 871265
Contact: Mrs. S. Stone
Established: 1979
Open: 9.30am to 5pm Monday to Friday.
Saturday by appointment
Craftworkers on site: 2
Crafts: Hand marbled paper. Lampbases,
candle lampshades, storage boxes,
albums
Prices: £2.95 to £95. Visa/Mastercard/
Access
Note: Workshops and demonstrations by
appointment

FISHERTON MILL DESIGN AND
CONTEMPORARY CRAFT EMPORIUM
108 Fisherton Street
Salisbury
Wiltshire
SP2 7QY
Tel: 01722 415121
Fax: 01722 415121
Contact: Leonie Nanassy/Michael Main
Established: 1994
Open: Monday to Saturday 9.30am to
5.30pm
Stock: High quality domestic pottery,
studio ceramics, wood, furniture, metal,
jewellery, textiles, glass, toys, leather as
well as fine art and sculpture. This
gallery shows work by over 250 artists
and designers. Commissions welcomed
Prices: From under £10 to over £5000
Note: Converted grain mill with
contemporary craft and design dramati-
cally displayed on three floors. In-house
restaurant and private car park

WINCHESTER STREET GALLERY
36 Winchester Street
Salisbury
Wiltshire
SP1 1HG
Tel: 01722 338235
Established: 1990
Open: Tuesday to Saturday 10.30am to
5pm
Stock: Paintings mostly by local artists.
Rugs, wall hangings, baskets, wooden
furniture and bowls, metal furniture,
pedestals, chimes, candlesticks. Sculpture, papier mâché, silver, glass jewellery,
handwoven waistcoats, scarves, painted
silk scarves, screens, cushions, glass and
ceramics

BELL STREET GALLERY
17 Bell Street
Shaftesbury
Dorset SP7 8AR
Tel: 01747 853540/01747 830748
Established: 1996
Open: 10am to 5pm
Stock: Contemporary work from the
Britsh Isles. All hand crafted of high
quality
Prices: £10 to £4,000
Note: Cafe downstairs with exhibition
space, a walled garden and licensed for
entertainment and alcohol. Seasonal
sculpture exhibitions. Disabled access

CRAFT STUDIO
4 Gold Hill Parade
Shaftesbury
Dorset SP7 8LY
Tel: 01747 854067
Contact: Chris Morphy
Open: 9.30am to 12.30pm and 1.30pm
to 5.30pm
Craftworkers on site: 1
Crafts: Silverware and jewellery
Prices: £3.50 to £500. MasterCard/Visa/
American Express
Note: Commissions, restoration and
repairs undertaken. Located at Gold Hill
beauty spot, museum, teashop & gift shop

LIBQA COURT CRAFT CENTRE
Fore Street
Sidmouth
Devon EX10 8AS
Tel: 01395 579090
Fax: 01395 579374
Contact: Charles and Susie Lodge
Established: 1993
Open: 9am to 5.30pm 7 days
Stock: Pottery, jewellery, stained glass,
wood turning, leather goods, découpage,
local artists, salt dough, cold cast bronze
figures all locally produced
Prices: 99p to £250. Access/Visa
Note: Pretty courtyard with other units
including silk flower shop, handmade
jumper shop, painted furniture/stencil
studio, antiques/militaria. Coffee shop
serving excellent coffee and teas

ROSE COTTAGE POTTERY
Harbourneford
South Brent
Devon TQ10 9DT
Tel: 01364 72550
Contact: Audrey Price
Established: 1982
Open: Drop in any time but phone first if
making a special visit
Craftworkers on site: 1
Crafts: Domestic and ornamental
earthenware decorated in lively colours
and designs
Prices: £5 to £80. Credit cards not
accepted
Note: Situated in attractive Dartmoor
hamlet. River Harbourne flows through
and under medieval clapper bridge. Rare
Victorian post box set in cottage wall.
Dartmoor beauty spot nearby. Member of
West Country Potters Association

RASHLEIGH POTTERY
Wheal Martyn China Clay Museum
Carthew
St. Austell
Cornwall PL26 8XG
Tel: 01726 850362
Contact: David Carew
Established: 1988
Open: 10am to 6pm
Craftworkers on site: 1
Crafts: Stoneware domesticware pottery
Prices: £1.95 to £50+. No credit cards
Note: The pottery is on site of the China
Clay Museum

NEW CRAFTSMAN
24 Fore Street
St. Ives
Cornwall TR26 1HE
Tel: 01736 795652
Contact: Mary Redgrave
Established: 1964
Open: Summer Monday to Friday 10am
to 6pm, Saturday 10am to 5pm. Winter
Monday to Saturday 10am to 5pm
Stock: High quality studio ceramics,
glass, wood, metal, jewellery as well as
paintings, sculpture and prints by artists
working in Cornwall.
Prices: From £5 to £7500. Major credit
cards accepted

SLOOP CRAFT CENTRE
St. Ives
Cornwall TR26 1LS
Tel: 01736 796051
Contact: Roy Harrison
Established: 1969
Open: 10am to 5pm Easter onwards
Crafts: Jewellery, leather goods, pottery,
paintings, stained glass, walking sticks,
wood carvings made on the premises
Prices: £5 to £500. Credit cards not
accepted
Note: Craftsmen working all the time

ST. IVES POTTERY GALLERY
1 Lower Fish Street
St. Ives
Cornwall TR26 1LT
Tel: 01736 794930
Contact: John Bedding
Established: 1991
Open: Monday to Saturday 9.30am to
5.30pm throughout the year
Stock: Studio pottery, stained glass
mirrors, raku ceramic jewellery

ENGLISH HURDLE
Curload, Stoke St. Gregory
Taunton
Somerset TA3 6JD
Tel: 01823 698418
Fax: 01823 698859
Contact: James Hector and Nigel Hector
Established: 1929
Craftworkers on site: 2+
Open: 9am to 5pm Monday to Friday.
Saturday 9am to 1pm
Crafts: Baskets, wattle fencing, plant
climbers, summerhouses, willow furniture
Prices: From £10 to £8000. Major credit
cards and Switch accepted
Note: Pleasant walk among growing
willow beds on the Somerset Levels.
Basket showroom also on premises

MAKERS
7a Bath Place
Taunton
Somerset TA1 4ER
Tel: 01823 251121
Contact: Solange
Established: 1984
Open: Monday to Saturday 9am to 5pm
Craftworkers on site: 10
Crafts: Jewellery, pottery, textiles, wood,
furniture, etchings, leather
Prices: £2 to £1,000. No credit cards
Note: One of England's first craft co-
operatives run on wholly equal terms

WILLOWS AND WETLANDS VISITOR CENTRE
Meare Green Court, Stoke St. Gregory
Taunton
Somerset TA3
Tel: 01823 490249
Fax: 01823 490814
Contact: Mrs. M.A. Coate
Established: Turn of century
Open: 9am to 5pm Monday to Saturday
Craftworkers on site: 10
Crafts: Willow baskets, furniture etc.
Prices: £2 to £100. Visa/Access/
MasterCard/Eurocard
Note: Working willow industry with
guided tours weekdays. Basket museum,
video of industry & environmental display

WREN POTTERY
The Village Arcade
10 Brook Street
Tavistock
Devon PL19 0HD
Tel: 01822 616896
Contact: Tim Farmer
Established: 1980
Open: 9am to 5pm Monday to Saturday
Crafts: Domestic and giftware pottery
Prices: 30p to £40. No credit cards

THE SILVER TREE
Fore Street
Tintagel
Cornwall PL34 0DA
Tel: 01840 770954
Contact: Angela Freke
Established: 1986
Open: High season open 7 days a week.
April to October Monday to Friday 10.15am
to 5pm, Saturday 10.15am to 1pm
Stock: Quality handmade silver jewellery,
wood turning, wooden jigsaws, paintings,
cards – all by local craftsmen
Prices: £2 to £500. Visa/MasterCard/
Eurocard
Note: Commissions undertaken

NICHOLAS CHANDLER FURNITURE MAKERS
"Woodpeckers"
Rackenford
Tiverton
Devon EX16 8ER
Tel: 01884 881380
Contact: Nicholas Chandler
Established: 1987
Open: 9am to 6pm Monday to Saturday
Craftworkers on site: 2
Crafts: Bookcases, boxes and bowls,
cabinets, cupboards, kitchens, dressers,
desks, dining tables and chairs, mirrors
magnificently made by hand
Prices: £15 to £75,000. Credit cards not
accepted

TIVERTON CRAFT CENTRE
1 Bridge Street
Tiverton
Devon EX16 5LY
Tel: 01884 258430
Contact: Mr. S.M. Chamberlin
Established: 1983
Open: 9am to 5.30pm Monday to
Saturday all year except January
Stock: Largest range of locally hand made
crafts in the South West. 150
craftspersons' work on display. Large
range of pottery
Prices: £1.25 to £800. Access/Visa/
American Express

BABBACOMBE POTTERY
Babbacombe Road
Torquay
Devon TQ1 3SY
Tel: 01803 323322
Fax: 01803 315444
Contact: Philip Laureston and Lynn Drake
Established: 1975
Open: Monday to Friday 9am to 5pm
Craftworkers on site: 4
Crafts: All stages of pottery manufacture can be seen, from raw clay to finished product. Locally made crafts
Prices: 25p to £50. Credit cards not accepted
Note: Visitors may try the potter's wheel and paint their own pottery. Portraits painted. Teashop/cafe and award winning gardens

LOTUS POTTERY
Stoke Gabriel
Totnes
South Devon TQ9 6SL
Tel: 01803 782303
Contact: Michael Skipwith
Established: 1957
Open: 9am to 12.30pm and 2pm to 5.30pm. Saturday by appointment
Crafts: Wood fired stoneware and porcelain, especially planters
Prices: £1 to £70. Credit cards not accepted
Note: Set in a garden/orchard in the heart of pretty Stoke Gabriel. Vice President Devon Guild of Craftsmen. Fellow Craft Potters Association

MARSHALL ARTS GALLERY
3 Warland
Totnes
South Devon
TQ9 5EL
Tel: 01803 863533
Contact: Chris Marshall
Established: 1986
Open: Summer: Monday to Saturday 10am to 1pm and 2pm to 5pm except Thursday and Saturday afternoons Winter: Tuesday, Wednesday, Friday 10am to 1pm and 2pm to 5pm. Saturday 10am to 1pm (closed Monday and Thursday)
Stock: Studio ceramics, wood turning, jewellery, painting and prints, glass
Note: Several exhibitions held each year

NUMBER SEVEN 'TOYS' FOR COLLECTORS
7 High Street
Totnes
South Devon
TQ9 5NN
Tel: 01803 866862
Contact: Vicki Wood
Established: 1994
Open: Fridays and Saturdays 10.15am to 5.30pm or by appointment
Craftworkers on site: 1
Crafts: Automata and moving 'toys' for adults
Prices: £7 to £1,000. Credit cards not accepted
Note: Occasional exhibitions of automata. No disabled access

GUILD OF TEN
19 Old Bridge Street
Truro
Cornwall
TR1 2AH
Tel: 01872 74681
Contact: G. Bamford
Established: 1980
Open: 9.30am to 1.30pm and 2pm to 5.30pm
Stock: Wood carving, ceramics, leather, jewellery, silk painting, scarves, dress designers, walking sticks, wood puzzles
Prices: £5 to £500.Visa/MasterCard/ Switch
Note: Co-operative run by members. Customers may meet craftsmen to discuss commissions and craftsmen sell direct to customers

NEW MILLS POTTERY
The Pottery
New Mills
Ladock
Truro
Cornwall
TR2 4NN
Tel: 01726 882209
Contact: John Davidson
Established: 1965
Open: 10am to 5pm Monday to Friday but advisable to phone first
Craftworkers on site: 1
Crafts: Handmade porcelain and stoneware. Individual pieces and domestic ware
Prices: £5 to £100. Credit cards not accepted

ARTY FACTS
The Platt
Wadebridge
Cornwall
PL27 7DS
Tel: 01208 812274
Contact: Fred and Mel Hardie
Established: 1989
Open: 9am to 5.30pm 6 days a week
Stock: Gifts and cards. Range of local and international craft work based on the natural world – wood, cotton, glass, pottery and more
Prices: 50p to £50. Access/Visa

PORTHILLY CRAFTS
near St. Michael's Church
Rock
near **Wadebridge**
Cornwall
PL27 6JX
Tel: 01208 863844
Fax: 01208 862624
Contact: Elizabeth Marshall
Open: Easter to end October 10am to 5pm. Closed Mondays. Out of season 10am to 5pm Friday to Sunday only. Other times by arrangement
Stock: Pottery, paintings, furniture, knitwear, cards, baskets, walking sticks, toys, mirrors, jewellery, glassware, candles and guidebooks. Plants and garden products
Note: Located at bottom of Porthilly Lane yards from the ancient church of St. Michael's. Large car park opposite

RIVERDEN FORGE
Roadwater
Watchet
Somerset
TA23 0QH
Tel: 01984 640648
Contact: Bill Poirrier
Established: 1965
Open: Weekdays 9am to 5pm. Weekends by appointment
Craftworkers on site: 1/2
Crafts: Blacksmithery. Stock items in forged steel, i.e. fire irons, brackets, candlesticks, door furniture, sculpture etc.
Prices: £2.50 to £450. Credit cards not accepted
Note: Working blacksmith. Commissions taken. One acre garden, 300 yards of river bank. Reasonable access for the disabled

WHITEHORSE POTTERY
Newtown
Westbury
Wiltshire
BA13 3EE
Tel: 01373 864772
Contact: Stephen Humm and Trevor Pictor
Established: 1979
Open: 10am to 5pm Monday to Saturday. Closed Sunday
Craftworkers on site: 2
Crafts: Handthrown terracotta gardenware and glazed domesticware
Prices: £2 to £200. Credit cards not accepted
Note: Workshop always open (during open times) for pottery making to be seen. Situated in attractive Victorian school on the road to Westbury White Horse. Car parking available

THE SQUARE GALLERY
Fore Street
Winkleigh
Devon
EH19 8HQ
Tel: 01837 83145
Contact: Sarah Flower
Established: 1996
Open: 10am to 5pm Tuesday, Thursday, Friday, Saturday. 10am to 2pm Wednesday. Closed Sundays and Mondays
Stock: High quality paintings (all media) traditional and contemporary, ceramics, sculpture wooden and bronze, etchings, photography, designer jewellery, metalwork, prints and cards
Prices: £15 to £1,500. Visa/Access/Mastercard
Note: Set in picturesque Devon village square in listed 18th century building. Free parking. Disabled access

NICK AGAR – WOODTURNER
Workshops 1 and 2
Buckland Abbey
Yelverton
Plymouth
Devon PL20 6EY
Tel: 01464 631485 (home)
Contact: Nick Agar
Established: 1991
Open: 10.30am to 4.30pm Friday to Wednesday. Closed Thursday
Craftworkers on site: 2
Crafts: High quality large bowls turned

from specialist woods. Sculptural and hollow forms a speciality – natural edged burrs etc. Light furniture work also produced

Prices: £5 to £500. Credit cards not accepted

Note: Prestigious clients include H.R.H. Prince of Wales and Aspreys. Gardens, restaurant, museum, walks, historic buildings (free entrance to craft work-shops only)

NICK DOUGLAS – POTTER
Buckland Abbey
Yelverton
Devon PL20 6EY
Tel: 01822 841220
Contact: Nick Douglas
Open: April 1st to October 1st 10.30am to 5pm daily except Thursday. 1st November to 31st March open weekends only 12.30pm to 5pm. Please telephone at other times
Craftworkers on site: 1
Crafts: Stoneware pottery
Prices: £4.50 to £160. Credit cards not accepted
Note: Member of the Devon Guild of Craftsmen. Buckland Abbey is a National Trust property. House, gardens, tearooms and events. Non NT members wishing only to visit the pottery should inform reception and they wil be given free admission

YELVERTON PAPERWEIGHT CENTRE
4 Buckland Terrace
Leg O'Mutton
Yelverton
Devon
PL20 6AD
Tel: 01822 854250
Contact: Mrs. Kay Bolster
Established: 1968
Open: Monday to Saturday 10am to 5pm. Please telephone for limited winter opening times
Stock: Glass paperweights, original oil paintings and prints by David Young and other established artists
Speciality: Selkirk glass and Perthshire paperweights
Prices: £3 to £2,000. Access/Visa cards are accepted
Note: Permanent exhibition of hundreds of glass paperweights "The Broughton Collection"

ROBINSONS
High Street
Queen Camel
near **Yeovil**
Somerset BA22 7NH
Tel: 01935 850609
Contact: Keith Robinson
Established: 1977
Open: 9am to 6pm Monday to Friday. Saturdays by appointment (or chance)
Craftworkers on site: 2
Crafts: Hand made furniture to authentic designs from stools and mirrors to large bookcases
Prices: £75 to £5,000. Credit cards not accepted
Note: Workshop can be viewed. Commissions taken. Individual attention with 40 years' experience. Set in beautiful village, easy parking. Wheelchair access can easily be arranged. Toilet on ground floor

HAWES CRAFT CENTRE AND TEAROOM
The Neukin
Hawes
near **Askrigg**
North Yorkshire
DL8 3RY
Tel: 01969 667727
Contact: Mr. and Mrs. Stocks
Established: 1984
Open: 10am to 5pm seven days. Closed Wednesday in winter
Stock: Pottery, jewellery, greeting cards, wooden toys, confectionery, soft toys, original paintings by owner/artist
Prices: 99p to £300+
Note: Tearoom with homemade fayre and fine views of the Dales

ARTLYNK GALLERY
14 Wednesday Market
Beverley
East Yorkshire
HU17 0DH
Tel: 01482 864902
Contact: Lynde A. Douthwaite
Established: 1976
Open: 10.30am to 5pm Tuesday to Saturday
Stock: Ceramics, glass, jewellery, wood, original prints, silk screen, etchings, linocuts, woodcuts, cards. Antique maps and prints
Speciality: Decorative and figurative ceramics
Prices: £5 to £2,000. Access/Visa/Mastercard accepted
Note: Twice yearly exhibitions by leading artists and craftsmen. Gallery is housed in a listed Georgian building in the environs of historic Beverley Minster. Teashop next door

BAILDON CRAFT SHOPS
Browgate
Baildon
Bradford
West Yorkshire
BD17 6BP
Tel: 01274 594946
Contact: Mrs. P. Ackroyd
Established: 1982
Open: 10am to 5pm Tuesday to Sunday
Stock: Silk/dried flowers, hand knitted wear, gifts for all occasions, picture frames and mounts, model railway and cars, kits, haberdashery, dolls houses and miniatures, pine furniture, candles, rugs, greeting cards, machine embroidery services (logos etc.) and much more
Speciality: Portrait artist and landscape artist. Commissions taken. Prices on request
Prices: 50p to £1,000. Credit cards not accepted
Note: Coffee shop serving hot and cold snacks and home baked goodies

PARK ROSE POTTERY & LEISURE PARK
Carnaby Covert Lane
Carnaby
Bridlington
East Yorkshire
YO15 3QF
Tel: 01262 602823
Fax: 01262 400202
Contact: Miss Jenny Pashby
Established: 1983
Open: 10am to 5pm (4.30pm winter)
Craftworkers on site: 50
Crafts: Pottery, giftware, planters, vases, lamp bases, bathroom accessories, general ceramic giftware lines. Hand decorated earthenware products
Prices: £2 to £30. Visa/Access
Note: Free pottery walkthrough with explanatory leaflet. Owl sanctuary, children's play area, cafeteria, gift shop. Disabled access

COXWOLD POTTERY
Coxwold
North Yorkshire YO6 4AA
Tel: 01347 868344
Contact: Peter Dick
Established: 1965
Open: 2pm to 5.30pm Tuesday to Friday and summer weekends
Craftworkers on site: 2
Crafts/Stock: Earthenware/stoneware pottery. Small items to special pieces of exhibition quality. Garden pots, candles, jewellery
Prices: £2.50 to £250. Credit cards not accepted
Note: Courtyard garden where plants and pots on sale – pleasant to sit and enjoy. No demonstrations but often work can be seen in progress. Please phone before making special visit

B & M TUFFNELL GLASSBLOWING
'Navenby', Westgate
Driffield
East Yorkshire YO25 7LJ
Tel: 01377 240745
Fax: 01377 240746
Contact: Bill and Martin Tuffnell
Established: 1978
Open: 8.30am to 5pm
Craftworkers on site: 9
Crafts/Stock: All types of glassware, industrial and decorative. Indian beads
Prices: 10p to £85. Visa/MasterCard/ Delta/Switch/JCB
Note: Demonstrations, talks about glass, its history, its uses, usually only to groups

COUNTRY FRAMES
25 Market Place
Driffield
East Yorkshire YO25 7AF
Tel: 01377 254082
Contact: Mr. N. Procter
Established: 1993
Open: 9.30am to 5pm Monday to Saturday
Stock: Pictures, picture framing, gifts, specialists in dog figurines and animal figurines, glassware, crystal and cards
Prices: 49p to £150. Visa/Access/Delta
Note: Friendly market town

CRAFT CUPBOARD
East Street, Kilham
Driffield
East Yorkshire YO25 0RE
Tel: 01262 420653
Contact: Ruth Kitcher
Established: 1990
Open: 10am to 5pm daily (except Monday and Wednesday). Open Bank Holiday Mondays
Stock: Hand crafted goods. Painting, embroidery, jewellery, pottery, découpage, glass, woodcraft, dough, knitwear, pressed flowers
Speciality: Hand crafted wood tables and picture framing
Prices: 99p to £120. No credit cards

PATRICK TITE CABINETMAKER & WOODCARVER
Galloway Lane
Driffield
East Yorkshire YO25 7LW
Tel: 01377 253689
Contact: Patrick Tite

Established: 1982
Open: Monday to Friday 9am to 5pm Saturday 9am to 12 noon
Craftworkers on site: 1
Crafts: Dining room furniture. One-off designs. Galloway Collection – English Oak
Note: Galloway Collection dining room suite can be seen in showroom

JASMINE DESIGNS CRAFT WORKSHOP
Marsh Lane
Beal
near **Goole**
Humberside
DN14 0SL
Tel: 01977 677417
Fax: 01977 677417
Contact: Angre Gannon
Established: 1996
Open: 9am to 5pm daily. Weekends 10am to 5pm
Craftworkers on site: 3
Crafts/Stock: Wide range of ceramic collectables such as ducks, pigs, cats and teddies. Uniquely designed handcrafted jewellery
Prices: £5 to £50. Visa
Note: Craft workshops held weekly in all aspects of decorating ceramics and monthly seminars. Country pub and restaurant next door

GALLERY '86, THE ART & CRAFT CENTRE
234 Freeman Street
Grimsby
North East Lincolnshire
DN32 9DR
Tel: 01472 351544
Contact: John and Win Taylor
Established: 1985
Open: 8.45am to 5pm. Closed Wednesdays at 12 noon
Stock: Basket weaving, découpage, jewellery findings, marquetry, glass staining, silk paints, stencils, embroidery, matchstick kits, flower pressing, full range of artists materials
Speciality: Highest quality bespoke picture framing
Prices: 35p to £30. Credit cards not accepted
Note: Local artists' paintings always available by Keith Baldock, Stephen Farrow, Carl R. Paul, Terri Dixon, Dale Mackay, George Odlin, Judith Corby. Speciality range of limited edition prints

JIM COOPER POTTERY & FORGE
103A Oldham Road
Ripponden
near **Halifax**
West Yorkshire HX6 4EB
Tel: 01422 822728
Contact: Jim Cooper
Established: 1984
Open: Please telephone for details
Craftworkers on site: 1
Stock/Crafts: Cast and thrown pottery.
Traditional hot forged iron work
Prices: £2.50 to £500. Credit cards not
accepted
Note: Turn of the century forge with
bellows and 6 hp open crank horizontal
Ruston paraffin engine driving line shaft

THROSTLE NEST GALLERY
Old Lindley, Holywell Green
Halifax
West Yorkshire HX4 9DF
Tel: 01422 374388
Contact: Pat Kaye
Established: 1986
Open: 10am to 5pm daily
Stock: Wide range of high quality British
craft. Changing exhibition on a regular
basis. Also paintings and selected greeting
cards
Speciality: Ceramics
Prices: £2.50 to £250. Credit cards not
accepted
Note: Potter on site producing mainly
domestic stoneware, but also large hand
built figures and animals. Wheelchair
accessible with assistance

GODFREY AND TWATT
7 Westminster Arcade
Parliament Street
Harrogate
North Yorkshire
HG1 2RN
Tel: 01423 525300
Contact: Alex Twatt
Established: 1985
Open: Monday to Saturday 10am to
5.30pm
Stock: High quality domestic pottery,
studio ceramics, glass, wood, jewellery as
well as other contemporary crafts and
original prints
Prices: From under £5 to around £1000.
Major credit cards are accepted
Note: Four exhibitions held each year.
Crafts Council selected gallery

KING STREET WORKSHOPS
King Street
Pateley Bridge
Harrogate
North Yorkshire
HG3 5LE
Contact: I. Simm
Established: 1992
Craftworkers on site: 3+
Open: 10am to 5pm Monday to Saturday
and occasional Sundays
Crafts: Glass, jewellery, ceramics,
terracotta ware outdoor pots/domestic/
designer and sculptural
Prices: £2 to £250. Credit cards not
accepted
Note: Glass blowing can be seen.
Nidderdale Museum on site. Disabled
access

BRIDGE MILL WORKSHOPS
Bridge Mill, St. George's Square
Hebden Bridge
West Yorkshire HX7 8ET
Tel: 01422 844559
Contact: J. Burgess, J. Cordingley, S.
Taylor
Established: 1989
Open: Generally Monday to Saturday and
some Sundays. Telephone before making
special journey
Craftworkers on site: 3
Crafts: Mill pottery – hand thrown
domestic stoneware. Unit 4 Enamelling –
enamel jewellery, pictures and clocks.
Sophie Taylor – papier mâché
Prices: £4 to £30
Note: Historical and picturesque small
Pennine town – spectacular countryside.
Crafts Council listed. No disabled access

HEBBLE END WORKS
Canal Towpath
Hebden Bridge
West Yorkshire HX7 6HJ
Contact: Rose Pratt
Established: 1992
Open: Easter to October 11am to 5pm
daily. October to Easter 12 noon to 5pm
weekends only. Other times by appoint-
ment
Crafts: Leatherwork, candles, models,
dolls' clothes, dolls' houses, beads,
sculptures
Prices: £1 to £50. Access/Barclaycard/
Visa
Note: Disabled access

THE ARK
113 North Road
Kirkburton
Huddersfield
West Yorkshire HD8 0RL
Tel: 01484 605055
Contact: Shirley Brinsley
Established: 1995
Open: Tuesday to Saturday 10am to 5pm
Stock: Jewellery, textiles, glass, metal, wood, domestic pottery
Prices: From £6 to £400. Major credit cards accepted
Note: Commissions undertaken

FREEGATE FOUNDRY
Freegate Mill, Cowling
Keighley
West Yorkshire BD22 0DJ
Tel: 01535 632723
Contact: Tom Ashby
Established: 1969
Open: 7.30am to 4.30pm Monday to Friday
Craftworkers on site: 8
Crafts: Decorative aluminium castings, Victorian reproduction garden furniture
Prices: £40 to £300. No credit cards
Note: Craft demonstrations

EUROPEAN CERAMICS
The Warehouse
Finkle Street
Knaresborough
North Yorkshire HG5 8AA
Tel: 01423 867401
Fax: 01483 867401
Contact: Maggie Barnes
Established: 1993
Open: Daily 11am to 5pm. Sunday 2pm to 5pm. Closed Thursdays
Stocks: Contemporary studio ceramics by leading makers from mainland Europe & UK – of particular interest to collectors
Prices: £5 to £3000. No credit cards at present but Yorkshire Artloan scheme available

COUNTRY CRAFTS
132 Harrogate Road, Chapel Allerton
Leeds
West Yorkshire LS7 4NZ
Tel: 01132 370597
Contact: Val Freer
Established: 1991
Open: 9.15am to 5.30pm Monday to Saturday
Stock: Patchwork, quilting, teddy bears,

oils, candles, hand painted glass, hand made cards, cushions, lace, mobiles and many smaller gifts
Speciality: Log cabin quilts, made to order and extensive range of beautiful Christmas crafts
Prices: 75p to £250. Access/Visa/ American Express
Note: Much of patchwork stock made on premises. Evening craft parties in Yorkshire area

DISPERSION '59'
Unit 0, Granary Wharf
Dark Arches, Leeds Canal Basin
Leeds
West Yorkshire LS1 4BR
Tel: 0113 2342925
Fax: 0113 2342925
Contact: Danny Johnston
Established: 1988
Open: 9.30am to 5.30pm 7 days a week
Stock: Ethnic clothing, home furnishings, brass and wooden figures (statues), incense, perfumed oils, jewellery
Prices: 25p to £1000+. Visa/Access
Note: Market stalls and street entertainers, bands and dancers, clowns etc. every Saturday and Sunday

THE CRAFT CENTRE & DESIGN GALLERY
The City Art Gallery
The Headrow
Leeds
West Yorkshire LS1 3AB
Tel: 01132 478240
Contact: Hayley Walker
Established: 1982
Open: Tuesday to Friday 10am to 5pm. Saturday 10am to 4pm
Stock: Contemporary jewellery, ceramics, textiles, limited edition prints, fine arts and crafts
Prices: £5 to £3,000. All credit cards except Diners accepted
Note: Regular exhibitions. Telephone for details. Leeds Art Gallery upstairs. Disabled access

CHANDLER GALLERY
Commercial Square
Leyburn
North Yorkshire DL8 5BP
Tel: 01969 623676
Contact: Charles and Daphne Chandler
Established: 1976
Open: 9.30am to 5pm. Closed Wednes-

day and Sunday. Closed February
Stock: Pottery, glass, jewellery, wood,
clocks. Mixture of gifts and crafts for all
tastes, pockets and ages
Prices: 3p to £1,500. Visa/Amex/
Mastercard
Note: Art gallery with permanent stock
and programme of solo and group
exhibitions. Disabled access to ground
floor (shop) only

LONGBARN ENTERPRISES
Low Mill, Bainbridge
Leyburn
North Yorkshire DL8 3EF
Tel: 01969 650416
Contact: Dr. Christopher Cole
Established: 1955
Open: Wednesdays 2pm to 5pm – July,
August and first half of September. Open
Bank Holidays but telephone in advance
Craftworkers on site: 1
Crafts: Dolls houses and accessories
Prices: £1.50 to £30. Credit cards not
accepted
Note: Fully furnished dolls houses. Restored
corn water mill. Small HO model railway.
Working model of a water mill

YORE MILL CRAFT SHOP
Aysgarth Falls
Leyburn
North Yorkshire DL8 3SR
Tel: 01969 663496
Contact: Valerie Woodley
Established: 1985
Open: Easter to end October 10am to
6pm. Mid-February to Easter and winter
weekends 11am to 5pm. Boxing Day to
New Year's Day 11am to 4.30pm.
Stock: Locally made wooden items.
British terracotta. Local pottery
Prices: £1 to £150. Barclaycard/Access
Note: Four rooms full of mainly British
gifts and crafts – something to suit
everyone's pocket. Mill-race teashop next
door. Two large car parks

GREEN MAN STONE
Hutton Hall Farm
Long Marston
North Yorkshire YO5 8LZ
Tel: 01904 738885
Contact: Gert van Hoff
Established: 1992
Open: 10am to 4pm weekends. Visitors
welcome weekdays, please 'phone first

Crafts: Stone sculpture. Hand-carved
medium sized stone ornaments
Speciality: Medieval themes reflecting
church carvings
Prices: £50 to £300. No credit cards
Note: Commissions undertaken from
house signs to pond features

DEERHOLME POTTERY
High Marishes
near **Malton**
North Yorkshire YO17 0UQ
Tel: 01653 668228
Contact: Sophie Hamilton
Established: 1991
Open: 10am to 6pm Monday to Friday
Craftworkers on site: 1
Crafts: Functional decorative stoneware pots
Prices: £4 to £75
Note: Member of Brigantia – Yorkshire
Moors and Coast Art and Craft Workers.
Disabled access

WOLDS SILVER
Rothay Cottage, Leppington
Malton
North Yorkshire YO17 9RL
Tel: 01653 658485
Contact: Vincent Ashworth
Established: 1986
Open: Please ring for details
Craftworkers on site: 1
Crafts: Silverware and jewellery
Note: Day and residential courses in
silversmithing. Excellent accommodation.
Leppington is small hamlet on edge of
Wolds in rich and varied rural country-
side. No facilities for disabled

**GORDON HODGSON HANDMADE
FURNITURE**
Cross Lane, Ingleby Cross
Northallerton
North Yorkshire DL6 3NQ
Tel: 01609 882414
Contact: Gordon Hodgson
Established: 1982
Open: Monday to Friday 9am to 5pm
Saturday 1pm to 5pm
Craftworkers on site: 1
Crafts: Individually made fine furniture
in variety of solid hardwoods. Each piece
carries a Bell trademark/day/item number
and signature
Prices: £10 to £5,000. Credit cards not
accepted
Note: Crafts Council listed.

AISLABY POTTERY
Michael & Wendy Salt
Aislaby Pottery
Aislaby
near **Pickering**
North Yorkshire
YO18 8PE
Tel: 01751 474128
Contact: Michael or Wendy Salt
Established: 1979
Open: Monday to Saturday 9.30am to
5.30pm. Sunday in summer 10am to
5.30pm. Monday and Sunday closed in
winter
Stock: Hand made, hand decorated, high
fired stoneware, including hand thrown
colourful oven to tableware. Lamps,
vases, bowls, tile panels and many one-
offs
Prices: £3 to £300. Credit cards not
accepted
Note: Located on south side of A170
Pickering to Helmsley Road, 2 miles west
of Pickering. Car parking for customers

GARDEN HOUSE POTTERY
The Green
Reeth
Richmond
North Yorkshire
DL11 6AB
Tel: 01748 884648
Contact: Ray and Jane Davies
Craftworkers on site: 2
Open: Easter to October (not Good
Friday) Monday to Saturday 10am to
5.30pm and other times by appointment.
Craft/stock: Architectural stoneware
items including number tiles and house
names as well some domestic pots, hazel
chairs and patchwork from local source
Prices: From under £5 to over £250.
Credit cards not accepted

JOY BENTLY SCULPTURE
East Windy Hall
Arkengarthdale Road
Reeth
Richmond
North Yorkshire
DL11 6QX
Tel: 01748 884316
Contact: Joy Bently
Craftworkers on site: 1
Open: First weekend of each month
Saturday and Sunday 2pm to 5pm, from
March to October. By appointment also

Craft: Figurative sculpture and portrait
work in bronze and resin bronze as well
as paintings and prints
Prices: Small sculptures from £150.
Commissioned portrait sculpture starts
from £1250 while the paintings are from
£50 upwards
Note: Portrait work can be commissioned

PHILIP BASTOW CABINETMAKER
Reeth Dales Crafts Centre
Silver Street
Reeth
Richmond
North Yorkshire
DL11 6SP
Tel: 01748 884555
Fax: 01748 884555
Contact: Philip Bastow
Established: 1982
Open: 9am to 5pm Monday to Friday,
9am to 12 noon Saturday, all year
Craftworkers on site: 2
Crafts: Furniture and wooden gifts
Credit cards accepted
Note: Specialising in the design and
manufacture of individual furniture
commissions

POTS 'N' PRESENTS
Anvil Square
Reeth
Richmond
North Yorkshire
DL11 6TE
Tel: 01748 884687/886377
Contact: Juliet and Martin Bearpark
Open: April to October 10am to 5.30pm
every day. March, November December
10am to 5pm weekends only
Stock: Exclusive range of stoneware
pottery. Over 50 diffferent handthrown
household items

STEFF'S MODELS
Anvil Square
Reeth
Richmond
North Yorkshire
DL11 6TB
Tel: 01748 884498
Fax: 01748 884334
Contact: Steff Ottevanger
Open: Monday to Friday 9am to 4pm all
year
Crafts: Hand-painted animal sculptures
and models

SWALEDALE POTTERY
Low Row
Richmond
North Yorkshire DL11 6TP
Tel: 01748 886377
Contact: Martin Bearpark
Open: 9am to 5pm Monday to Saturday
1st March to 31st October
Crafts: Traditional handthrown stoneware
pottery
Note: Visitors welcome to watch potter at
work. Finished items displayed at
workshop and at Pots 'n' Presents in
Reeth

ISLAND HERITAGE
Pott Hall Farm
Healey, Masham
near **Ripon**
North Yorkshire HG4 4LT
Tel: 01765 689651
Fax: 01765 689350
Contact: Penelope Webb
Established: 1982
Open: 10am to 6pm
Craftworkers on site: 4
Crafts: Clothes and accessories made
from the natural undyed fleece of rare
breeds of British sheep
Prices: £3 to £180
Note: Weaving, spinning, tailoring
demonstrations. Rare breed sheep, pigs,
goats, chickens and the Old Sheep Dip
picnic garden for customers. Entrance
free. 2.5 miles past Healey above
Leighton reservoir

MORTHEN CRAFT WORKSHOP
Springvale Farm, Morthen Lane
Wickersley
Rotherham
South Yorkshire S66 9JQ
Tel: 01709 547346
Contact: Mr. Brian Scott
Established: 1985
Open: 10am to 5pm daily
Craftworkers on site: 2
Crafts: Lapidary, jewellery, herbs,
cosmetics, floral art, Zimbabwe Shona
sculpture
Prices: £2 to £5,000. Credit cards not
accepted
Note: Open all year. Wide selection of
interesting demonstrations. Daytime and
evening group talks with refreshments –
please telephone for availability.
Teashop, gardens, herb plants, parking

WINGHAM WOOLWORK
70 Main Street
Wentworth
Rotherham
South Yorkshire
S62 7BR
Tel: 01226 742926
Fax: 01226 741166
Contact: Ruth Gough
Open: Saturday 11.30am to 5.30pm.
Sunday 11.30am to 5.30pm. Monday
1.30pm to 5.30pm
Craftworkers on site: 2
Crafts: Handspun knitwear/yarns.
Specialists in production of fibre blends
for handspinners
Prices: Various. Access/Visa accepted
Note: Teachers of handspinning

ANKARET CRESSWELL
Wykeham
Scarborough
North Yorkshire YO13 9QB
Tel: 01723 864406
Fax: 01723 864041
Contact: Justin Terry
Established: 1986
Open: Monday to Friday 9.30am to
12.30pm and 2pm to 5pm
Craftworkers on site: 2
Crafts: Wool fabric weaving and
tailoring. Clothing accessories
Prices: £60 to £260. Visa/MasterCard/
Switch/Delta
Note: Working floor loom sometimes in
action. Another general craft shop in
same location with teas etc. Easy parking.
Serious callers please telephone first

ONE WORLD
18 Victoria Road
Scarborough
North Yorkshire YO11 1SD
Tel: 01723 360754
Contact: Kerry Ward
Established: 1992
Open: 10am to 5pm Monday to Saturday
Stock: Crafts from the third world. Goods
made locally. Ethnic clothing and
jewellery
Prices: 20p to £50. Credit cards not
accepted
Note: One World is first and foremost a
charity shop enabling Scarborough
Council for Voluntary Service to provide
rent free accommodation to Home and
Dry, a project for the young homeless

OAKLEAF CABINETMAKERS
20 Fieldside, Crowle
near **Scunthorpe**
North Lincolnshire
DN17 4HL
Tel: 01724 711027
Fax: 01724 710018
Contact: Stephen Moore
Established: 1982
Open: Monday to Friday 9am to 5pm.
Saturday 9am to 4.30pm. Other times
visitors welcome by appointment
Craftworkers on site: 3
Crafts: Reproduction handmade quality
furniture
Prices: Visa/Acccess accepted
Note: Furniture, kitchens, bedrooms on
display in showroom. Commissions
undertaken. Private and commercial
clients welcome

DALESMADE CENTRE
Watershed Mill Visitor Centre
Langcliffe Road
Settle
North Yorkshire BD24 9LR
Tel: 01729 825111
Fax: 01729 825202
Contact: Lynne Ridgway
Established: 1991
Open: Monday to Saturday from 10am.
Sundays from 11am. Closed Christmas
Day and Easter Sunday
Stock: Pottery, leatherwork, paintings,
prints, photography, clothes, woodwork,
preserves, slippers, stained glass, flowers,
toys, bags, jewellery
Speciality: Quality products from over 40
Yorkshire Dales' craftworkers
Prices: 99p to £100. Visa/Delta/
MasterCard
Note: Located in converted 19th century
textile mill in the Dales with Hector's
Coffee Shop, Rock & Fossil Shop, The
Edinburgh Woollen Mill Country Store,
art exhibitions and craft demonstrations
(seasonal). Disabled access and facilities
throughout

DÉCOUPAGE CRAFT AND TEACHING CENTRE
851 Chesterfield Road
Sheffield
South Yorkshire S8 0SQ
Tel: 01142 746194
Contact: Muriel Stanley
Established: 1992
Open: Daily 9am to 4.30pm. Saturday
9am to 1pm. Closed Sunday and Monday
Stock: Prints, découpage accessories,
specialised 3D pictures (to order).
Pergamano – parchment craft
Speciality: Craft teaching and framing
Prices: 35p to £700. Credit cards not
accepted

FEATHERSTONE & DURBER
The Barn, Limestone Cottage Farm
Limestone Cottage Lane, Wadsley Bdg
Sheffield
South Yorkshire S6 1NJ
Tel: 01142 853775
Fax: 01142 853775
Contact: Sarah Featherstone and Jeff
Durber
Established: 1993
Open: 10am to 5pm but telephone for
appointment
Craftworkers on site: 2
Crafts: Metalwork, jewellery and pocket
knives
Prices: £20 to £200. Credit cards not
accepted

THE CRAFT CENTRE
Rother Valley Country Park
Mansfield Road, Wales Bar
Sheffield
South Yorkshire S31 8PE
Tel: 01142 511396
Fax: 01142 481251
Contact: Jenny
Established: 1988
Open: 11am to 5pm daily. Certain crafts
may be closed Monday or Tuesday
Craftworkers on site: 4
Crafts: Handmade pottery, bead
jewellery, floral arrangements
Prices: 50p to £100. Credit cards not
accepted
Note: Set in parkland. Cafe. Visitors can
see craft works in progress. Disabled
access to pottery but jewellery and
flowers are upstairs and no specialised
arrangements made

THE SHEILA SAWYER DÉCOUPAGE CENTRE
Craft Workshop, 4 Orchard Square
Shopping Precinct, Fargate
Sheffield
South Yorkshire
S1 2FB
Tel: 01142 796477

Contact: S. and R. Sawyer
Established: 1991
Open: 10am to 5pm Monday to Saturday
Crafts: Everything for elevated découpage. Frame, prints, silicone, glaze
Speciality: Finished work by Sheila Sawyer
Prices: £10 to £400. Visa/Switch
Note: Sheila works in shop every day. Lessons Monday to Friday. Seminars on some Saturdays

GEMINI STUDIOS
51A Main Street, Grassington
Skipton
North Yorkshire BD23 5AA
Tel: 01756 752605
Contact: Sheila Denby
Established: 1984
Open: 10.30am to 5.30pm
Craftworkers on site: 2
Crafts: Sterling silver/9ct gold jewellery and small artefacts (boxes, dishes)
Prices: £2 to £495. Credit cards not accepted
Note: Jewellery can be seen being made on premises. Katie Denby is Viking Museum's (York) Exhibition Silversmith attending their two exhibitions each year

OLD HALL WORKSHOP
Carlton Husthwaite
Thirsk
North Yorkshire YO7 2BQ
Tel: 01845 501359
Contact: Malcolm Pipes
Established: 1981
Open: 9am to 5pm daily
Craftworkers on site: 2
Crafts: Candlestick holders, trolleys, nests of tables, dining tables, oak book troughs, ashtrays, chairs, napkin rings and holders, stools etc. Machine knitwear. Work made to order

TRESKE
Station Maltings
Thirsk
North Yorkshire YO7 4NY
Tel: 01845 522770
Fax: 01845 522692
Contact: Nicola Fobbs
Established: 1973
Open: 7 days – 10am to 5pm summer and 11am till dusk winter
Craftworkers on site: 20
Crafts: Solid wood furniture and gifts

Prices: £1 to £1,000 plus. Most credit cards accepted. Tour £2.50, £1 child
Note: Exhibition on wood, tearooms, giftshop – all in our visitor centre. Open to group bookings at reduced rates

THE CLOGSHOP
27 Queen Street, Horbury
Wakefield
West Yorkshire WF4 6LP
Tel: 01924 281997
Contact: Trefor Owen and Liz Higgs
Established: 1995
Open: Normally 9.30am to 5pm Monday to Saturday
Craftworkers on site: 2
Crafts/Stock: Clogs, leather, ceramics, clothing, prints, crystal and glass etc
Prices: £7.50 to £100. Credit cards not accepted
Note: Permanent clog making (shop and workshop combined)

ARTEFACTS
159 Church Street
Whitby
North Yorkshire YO22 4AS
Tel: 01947 820682
Contact: Amanda and Steven Smith
Established: 1991
Open: Monday to Saturday 10am to 5pm, Sunday 11am to 4pm (closed Wednesdays). Winter Friday, Saturday, Sunday Monday
Stock: Needlework specialist design centre. Wooden craft and pottery, needlework threads, fabrics, beads, patterns etc.
Prices: 50p to £150. Credit cards not accepted
Note: Demonstrations given. Located in 14th century timber frame building in beautiful seafaring town

ORCHARD CRAFTS
Egton Bridge
near **Whitby**
North Yorkshire YO21 1XE
Tel: 01947 895483
Contact: Mrs. Ann Woollin
Established: 1990
Open: 10am to 6pm all year. Closed Sundays. In summer may be open to 9pm
Stock: Fabrics. Local wooden toys and soft toys. Wrought ironwork. Turned woodwork. Paintings, cards and gifts
Speciality: Cushions, bags, jackets, mats,
Note: Shop in beautiful setting

PENNY HEDGE GALLERY
135 Church Street
Whitby
North Yorkshire YO22 4DE
Tel: 01947 821310
Contact: Geoff Carnes
Established: 1993
Open: Easter to end October 10am to
5.30pm daily. Winter 10am to 5pm
Thursday to Monday inclusive
Stock: Paintings by local and British artists
Prices: £40 to £500. No credit cards

THE GROSMONT GALLERY
The Old Workshop, Grosmont
near **Whitby**
North Yorkshire YO22 5QE
Tel: 01947 895442
Contact: Mrs. A. Blackwell
Established: 1978
Open: 12 noon to 5pm Weekdays and
Sunday. 10am to 5pm Saturdays
Stock: Ceramics, paintings, prints, ready
made frames, artists' materials
Speciality: Original contemporary work
Prices: £5 to £500. Visa/Access
Note: Studio of resident artists Paul
Blackwell and Anne Thornhill – range of
exclusive prints available. Extensive
display of work housed in building of
great character and friendly atmosphere

THE STAINED GLASS WORKSHOP
The Studio, High Street, Lythe
near **Whitby**
North Yorkshire YO21 3RT
Tel: 01947 810101/893246
Contact: Alan Davis
Established: 1990
Open: 9.30am to 4.30pm Monday to
Saturday
Craftworkers on site: 2
Crafts: Repair & restoration of stained glass
Prices: £100 plus

DEFINITELY DIFFERENT
4 Pinders Court, Stonegate Walk
York
North Yorkshire YO1 2QQ
Tel: 01904 631856
Contact: Emma Moon
Established: 1993
Open: 10am to 5pm – longer during
holiday periods and Christmas
Stock: A full range of British gifts and
small craft work
Prices: £1 to £100

JACQUELINE JAMES HANDWOVEN RUGS
4 Rosslyn Street
York
North Yorkshire YO3 6LG
Tel: 01904 621381
Contact: Jacqueline James
Established: 1989
Open: By appointment
Craftworkers on site: 1
Crafts: Individually designed handwoven
rugs and wall hangings
Prices: £150 to £500
Note: Commissions accepted. Major
commissions include weaving for
Westminster Abbey and York Minster

PYRAMID GALLERY
43 Stonegate
York
North Yorkshire YO1 2AW
Tel: 01904 641187
Contact: Elaine Brett
Established: 1982
Open: Monday to Saturday 10am to 5pm.
Sundays in peak seasons 11am to 4pm
Stock: High quality domestic pottery,
studio ceramics, glass, wood, metal,
jewellery, textiles as well as original prints
Prices: From £15 to £2500, all major
credit cards accepted plus Switch

ROBERT FEATHER JEWELLERY GALLERY
10 Gillygate
York
North Yorkshire YO3 7EQ
Tel: 01904 632025
Open: Monday to Saturday 9.30am to
5.30pm
Stock: High quality contemporary jewellery
plus selection of ceramics and glass
Prices: £25 up to around £1000. All major
cards accepted plus Switch
Note: Situated near to City Art Gallery

THE POTTERY
Appleton-le-Moors
York
North Yorkshire YO6 6TE
Tel: 01751 417514
Contact: James R. Brooke
Established: 1968
Open: Every day 10am till dusk
Craftworkers on site: 1
Crafts: Slip decorated hand thrown
earthenware
Prices: £1.50 to £25. No credit cards
Note: Reasonable access for disabled

AMALGAM
3 Barnes High Street
Barnes SW13 9LB
Tel: 0181 878 1279
Contact: Tim Boon
Established: 1973
Open: Tuesday to Saturday 10am to
1.30pm and 2.30pm to 6pm
Stock: High quality ceramic container
forms by new and established makers.
Also blown glass, jewellery and prints
Prices: From under £10 to over £1000
Major credit cards accepted
Note: One of the leading London
ceramics galleries

PEARL DOT LTD
2 Roman Way
Barnsbury N7 8XG
Tel: 0171 609 3169
Fax: 0171 607 3904
Contact: James Summers
Established: 1976
Open: 8am to 6pm
Craftworkers on site: 7
Crafts: Furniture – one-offs, small batch,
prototypes, etc.
Prices: £50 to £50,000. Credit cards not
accepted

CHELSEA POTTERY
Ebury Mews
Belgravia SW1W 9NX
Tel: 0171 259 0164
Contact: Richard Dennison
Established: 1952
Open: Monday to Friday 9.30am to
5.30pm
Crafts: Hand made hand decorated
pottery
Prices: £5 to £175
Note: Disabled access

GALERIE SINGLETON
40 Theobalds Road
Bloomsbury WC1X 8NW
Tel: 0171 831 6928
Fax: 0171 831 6928
Contact: Jeremy Banks
Established: 1992
Open: Monday to Friday 10.30am to
6.30pm. Saturday 11am to 5pm
Stock: Mirrors, glassware, clocks,
ceramics, silver/gold jewellery, lighting,
furniture, handmade lamps, Indian

imports as well as paintings
Speciality: Copper/verdigris clocks
Prices: £5 to £500. MasterCard/Visa/
Delta/Switch/Electron/Eurocard

KEW BRIDGE FORGE
Green Dragon Lane
Brentford
Middlesex TW8 0EN
Tel: 0181 568 4459
Fax: 0181 569 9978
Contact: Kaspar Swankey
Established: 1992
Open: 10am to 6pm
Craftworkers on site: 4
Crafts: Furniture and architectural
ironwork
Note: Specialising in one-off contempo-
rary commissions. Disabled access. Kew
Bridge Steam Museum nearby
Note: Disabled access

THE LOCK SHOP
1st Floor, 9 Chalk Farm Road
opp. Camden Lock
Camden NW1 8AA
Tel: 0171 485 3450
Contact: Cathy Goodhead
Established: 1973
Open: 10am to 6pm Monday to
Saturday. 10.30am to 7pm Sunday
Stock: Variety of gift items including seals
and rubber stamps, stationery, pottery,
novelty teapots, games
Speciality: Windchimes
Prices: £1.99 to £150. MasterCard/Visa/
Amex/Diners Club
Note: Located opposite Europe's biggest
street market which is open weekends.
Short walk to Regent's Canal & London Zoo

THE TRADING POST
40 Middle Yard, Camden Lock Place
Camden NW1 8AF
Tel: 0171 284 2089
Established: 1992
Contact: Scott Taylor
Open: 11am to 5.30pm Wednesday to
Friday. 10am to 6pm weekends
Stock: Wide range of genuine native
American Indian arts and crafts
Speciality: Sand paintings, silver and
turquoise jewellery, dream catchers,
smudge, fetish necklaces
Prices: £2 to £500

ARGENTA
82 Fulham Road
Chelsea SW3 6HR
Tel: 0171 584 1841
Fax: 0171 584 3119
Contact: K. Jacobsen
Established: 1967
Open: Monday to Friday 9.30am to
5.30pm. Saturday 9.30am to 5pm
Stock: Contemporary jewellery from over
100 artists from Britain and abroad
Prices: From under £10 to over £5,000.
Visa/MasterCard/Amex accepted
Note: Commissions welcomed

FRIVOLI
7a Devonshire Road
Chiswick W4 2EU
Tel: 0181 742 3255
Fax: 0181 994 7372
Contact: Hazel Peiser
Established: 1991
Open: Monday to Saturday 10am to 6pm
Stock: Top quality crafts from clothing to
furniture, toys and ceramics, glass,
woodcarving and turning etc.
Prices: £2 to £1,000. Credit cards
accepted
Note: Themed exhibitions held each year

BAGGY BOTTOM DESIGNS
Unit 38 Cornwell House
21 Clerkenwell Green
Clerkenwell EC1R 0DP
Tel: 0171 336 6837
Contact: Rebecca Skeels
Established: 1994
Open: 9am to 5pm Monday to Friday.
Other times by appointment
Craftworkers on site: 1
Crafts: Individually designed and
handcrafted one-off pieces of silver
contemporary jewellery
Prices: £5 to £50. No credit cards
Note: Work made to order. Group shows
held twice a year with other local
craftspeople. Other workshops include
textiles, ceramics, bookbinding etc. No
disabled access as lots of stairs.

LESLEY CRAZE GALLERY
34 Clerkenwell Green
Clerkenwell
EC1R 0DU
Tel: 0171 608 0393
Fax: 0171 608 0393
Contact: Lesley Craze

Open: Monday to Saturday 10am to 5.30pm
Stock: Precious and non-precious
contemporary jewellery. Innovative and
exciting designs using materials as diverse
as papier mâché, acrylic, silver and 18ct
white and yellow gold
Prices: £10 to £1,000+. Visa/MasterCard/
Amex/Switch
Note: Commissions undertaken. Crafts
Council selected gallery. Situated within a
pleasant green in the borough of Islington
- a thriving & fascinating area of London

BEST OF BRITISH
24 Thomas Neal's
Earlham Street
Covent Garden WC2H 9LD
Fax: 0171 379 4097
Contact: Mrs. Bruton
Established: 1988
Open: 10.30am to 7pm March to
December. 10.30am to 6pm January to
March. Occasional Sundays 1pm to 5pm
Stock: Pottery, knitwear, pictures, toys,
bags
Prices: 50p+. Visa/MasterCard/Amex/
Diners/JCB
Note: Mail order available. Located in
covered shopping precinct opposite cafe

FALCONER
11 Topsfield Parade
Crouch End N8 8PR
Tel: 0181 348 6228
Contact: Sarah Falconer
Established: 1977
Open: Monday to Saturday 9am to 6pm.
Sunday 10am to 6pm.
Christmas – Monday to Saturday 9am to
8pm and Sunday 10am to 6pm
Stock: Ceramics, silver, glassware, textiles
and toys
Prices: 45p to £300. Switch/Access/
Barclaycard
Open: Many interesting tea and coffee
shops in villagey Crouch End

ART WITH GLASS
Stained Glass Workshop
Haynes Lane
off Westow Street
Crystal Palace SE19 3RW
Tel: 0181 771 6845
Contact: Sandra Palmer or Vincent Reid
Open: 9.30am to 6pm Tuesday to Friday.
10am to 5pm Saturday
Craftworkers on site: 2

Crafts: Stained glass – panels, lights, mirrors, gifts
Prices: £1.99 to £1,000. Credit cards not accepted
Note: Items made to individual's specifications. Disabled access

MONTEZUMAS
9 Oak Road
Ealing W5 3SS
Tel: 0181 579 6293
Fax: 0181 566 2758
Contact: Lucy Wayne/Moira Wiggins
Open: 10am to 6.30pm Monday to Saturday. 10am to 7pm Thursday
Stock: Ethnic items including carvings, Mexican tin ware, mirrors, glasses, throws, wrought iron
Prices: 50p to £1,000. Visa/MasterCard/Amex
Note: Vast range of stock from furniture to jewellery

EVER PRESENCE
12 The Quadrant
Manor Park Crescent
Edgware
Middlesex HA8 7LU
Tel: 0181 951 4000
Mobile: 0378 435931
Contact: Marilyn Taylor
Established: 1995
Open: Wednesdays and Fridays 10am to 4pm. Saturdays 9.30am to 4.30. Sundays 10.30am to 2pm
Crafts/stock: Dough dollies, glassware, petal porcelain, canalware, knitted items, herb boxes, native American Indian art and jewellery, terracotta
Speciality: Cross-stitch, Canalware, rubber stamps
Prices: £2 to £40. Credit cards not accepted
Note: Craft classes and demonstrations held. Workshops in unusual subjects, e.g. tarot, astrology, American Indian. Crystal, tarot and incense also for sale. Readings available

ARCHIPELIGO
Unit 38
Pennybank Chambers
33-35 St Johns Square
Finsbury
EC1M 4DS
Contact: Doreen Gittens
Established: 1996

Open: 9am to 6pm
Craftworkers on site: 1
Crafts: Handwoven silk scarves, wraps and jackets
Prices: £50 to £200. Credit cards not accepted
Note: Disabled access

YEMANJA A WINDOW TO AFRIKA
7 Stroud Green Road
Finsbury Park
N4 2DQ
Tel: 0171 281 8668
Contact: Steve Henry
Established: 1993
Open: 10.30am to 7pm
Stock: African arts and crafts including prints, paintings, jewellery, fabrics, greeting cards, drums, masks
Speciality: Books (Afrocentric)
Prices: £1 to £250. All major credit cards accepted
Note: Most unusual display of goods from all over Africa

ART IN IRON
195 Townmead Road
Fulham SW6 2QQ
Tel: 0171 384 3404
Fax: 0171 384 3404 (call first)
Established: 1993
Open: 10am to 6pm Monday to Friday. 11am to 5pm Saturday and Sunday
Crafts: Iron beds and furnishings
Prices: £385 to £800. Visa/Access/MasterCard
Note: Working factory on weekdays – see iron beds being made, welding, grinding, painting

STAINED GLASS INTERIORS
12 Dawes Road
Fulham SW6 7EN
Tel: 0171 385 5888
Contact: "Sherif" Amin
Established: 1995
Open: 9am to 5pm Monday to Saturday. 9am to 1.30pm Saturday
Craftworkers on site: 1
Crafts/stock: Stained glass windows, hand painted fired and leaded. All stained glass tools, materials and glass
Prices: 30p to £65.50. Credit cards not accepted
Note: Working studio with regular exhibitions and trips to workshops abroad etc. Disabled access to ground floor shop

QUETZAL
No 1 The Market
Greenwich
SE10 9HZ
Tel: 0181 305 0429
Contact: Simon Pollock
Established: 1989
Open: Wednesday to Friday 12 noon to
5pm. Weekends 11am to 5.30pm
Stock: Clothing, accessories and crafts
from Guatemala, Ecuador, Peru, Thailand
and Indonesia
Speciality: Handwoven and dyed cotton
clothing
Prices: £1 to £60. Most credit cards
accepted

"STS"/STOJANKA STRUGAR
73 Eton Avenue
Hampstead
NW3 3EU
Tel/Fax: 0171 483 1867
Contact: Stojanka Strugar
Open: 11am to 5pm
Stock: Hand-woven tapestry and unique
textiles
Prices: £100 to £2,000. Credit cards not
accepted
Note: Tapestries made with natural
materials: wool, silk and cotton

ARTEMIDORUS
27b Half Moon Lane
Herne Hill
London
SE24 9JU
Tel: 0171 737 7747
Contact: Amanda Walbank
Established: 1991
Open: Tuesday to Friday 10.30am to
7pm. Saturday 10.30am to 6pm, Sunday/
Monday throughout December
Stock: Ceramics and glass, jewellery,
metalwork, silk and textiles, wood, toys,
aromatherapy, glasses, beakers, vases,
jugs and bowls, mirrors, clocks, ties and
scarves, knitwear, cushions and throws,
lampshades and candlesticks
Speciality: Nearly everything is made by
hand. Selected contemporary British
artists, makers and crafts people create
over 95% of the items on display
Prices: From under £1 to over £500.
Credit cards accepted
Note: Exhibitions held in gallery upstairs
including etchings, watercolours, textile
works, ceramics etc.

MOSAIC WORKSHOP
Unit B, 443/9 Holloway Road
Holloway
N7 6LJ
Tel: 0171 263 2997
Fax: 0171 263 2997
Contact: James Postgate
Established: 1987
Open: Tuesday to Friday 10am to 6pm
Craftworkers on site: 5
Crafts: Mosaic of all sorts and mosaic
making materials
Prices: £1 to £1,000. Credit cards not
accepted
Note: Mosaic courses held occasionally

ARIA
133 Upper Street
Islington
N1 1QP
Tel: 0171 226 1021
Fax: 0171 704 6333
Contact: Pushpa Gulhane
Established: 1989
Open: 10am to 7pm Monday to Friday.
10am to 6.30pm Saturday. 12 noon to
5pm Sunday
Note: Also at 295 Upper Street, Islington
N1 2TU. Tel: 0171 704 1999

CRAFTS COUNCIL GALLERY SHOP
44A Pentonville Road
Islington
N1 9BY
Tel: 0171 806 2559
Fax: 0171 837 6891
Contact: Clare Beck
Established: 1991
Open: Tuesday to Saturday 11am to 6pm
Sundays 2pm to 6pm
Stock: Ceramics, glass, wall hung textiles,
wearables, furniture, jewellery
Speciality: Craft publications and high
quality UK crafts
Prices: £12 to £200. Visa/Access/Amex
Note: Within Crafts Council head
quarters. Teashop (cafe) and workshops

LUCIANA IZZI DESIGNS
169 Liverpool Road
Islington
N1 0RF
Tel: 0171 359 7968/833 0859
Contact: Luciana Izzi
Established: 1995
Open: 10am to 8pm
Craftworkers on site: 2

Crafts: Mosaics, handpainted glasses, handmade mirrors, photo frames, candleholders, glass boxes, handpainted bottles
Prices: £2.20 to £20. Credit cards not accepted

PACO MUÑOZ DESIGN
Flat 2
1 Highbury Terrace
Islington
N5 1UP
Tel: 0171 226 4910
Fax: 0171 226 4910
Contact: Paco Muñoz
Established: 1989
Open: Most days but please 'phone first
Craftworkers on site: 1
Crafts: Enamelled silver jewellery
Note: Commissions undertaken, repairs to enamelling and bespoke design service

THE GLASSHOUSE
21 St. Alban's Place
Islington Green
N1 0NX
Tel: 0171 359 8162
Fax: 0171 359 9485
Contact: Ann Morgan
Established: 1969
Open: Tuesday to Friday 10am to 6pm. Saturday 10am to 5pm
Crafts/stock: Studio glass: vases, candlesticks, bowls, napkin rings and 'one-off' pieces by Annette Meech, Fleur Tookey, David Taylor and Christopher Williams
Prices: £4.50 to £2,000. Visa/ MasterCard/Amex/Diners
Note: Glass blowing workshop open to visitors all year round Tuesday to Friday 10am to 5.30pm (closed for lunch 1pm-2pm). Disabled access to ground floor gallery only

THE PEA-GREEN POTTERY
Unit 3 Peabody Yard
Greenman Street
Islington
N1 8SE
Tel: 0171 359 5646/0171 226 8464 (home)
Contact: Andrea Peters
Established: 1996
Open: Monday to Friday 11am to 4pm
Craftworkers on site: 2
Crafts: Pottery/ceramics by Andrea Peters

and Stanley Corby. Also paintings and sculpture by Andrea Peters
Prices: £10 to £1,000. Credit cards not accepted
Note: Both artists work in traditional methods using their own contemporary designs. Demonstrations held and commissions undertaken. Located in Victorian cobbled yard with a distinct "Dickens-like" feel

VICTORIA SMITH & JANE CROCKETT
4 Clements Yard
Iliffe Street
Kennington SE17 3LJ
Tel: 0171 582 8285
Fax: 0171 582 4489
Contact: Victoria Smith
Established: 1994
Open: 11am to 7pm Monday to Friday
Craftworkers on site: 2
Crafts: Jewellery
Prices: £10 to £200. Credit cards not accepted
Note: Disabled access. We also sell at Greenwich Craft Market

ENID LAWSON GALLERY
36A Kensington Church Street
Kensington W8 4BX
Tel: 0171 937 8444
Fax: 0171 938 4786
Open: 10.30am to 6pm
Stock: Fine studio ceramics and contemporary fine art
Prices: From under £20 to over £1,000. Credit cards

BILLY ZERO NICHOLAS
Unit 45, Kingsgate Workshops
110-116 Kingsgate Road
Kilburn NW6 2JG
Tel: 0171 328 8286
Fax: 0171 328 7878
Contact: Billy Nicholas
Established: 1983
Open: 1pm to 8pm
Craftworkers on site: 1
Crafts: Masks (papier mâché, resin), brooches, jewellery, sculpture
Prices: £3.50 to £200. No credit cards
Note: Kingsgate Workshops has 50 units of artists and craftspeople. Please phone for details of workshop open days. Full programme of exhibitions held in Kingsgate Gallery on premises. Disabled access to gallery only

A.D.L. TACK REPAIRS
164 Forest Road
Leyton
E11 1LF
Tel: 0181 539 8100
Contact: Adrian Liddle
Established: 1990
Open: Monday to Friday 2pm to 6pm.
Saturday 10am to 6pm. Sunday 11am to
4pm
Craftworkers on site: 1
Crafts: Leatherwork for saddlery trade
and other leather items to order
Prices: £1 to £500. Visa/MasterCard
accepted
Note: Mainly deal with equine trade but
will make or repair anything to do with
leather except shoes and boots. No
disabled access

FITCH'S ARK
6 Clifton Road
Little Venice
W9 1SS
Tel: 0171 266 0202
Fax: 0171 266 0060
Contact: Susannah Fellows
Established: 1994
Open: Monday to Saturday 11am to 7pm.
Sunday 11am to 3pm
Stock: High quality domestic pottery,
studio ceramics, wood, furniture, metal
jewellery, textiles, glass, and toys
Prices: From under £5 to over £5000
Major credit cards accepted
Note: Unusual gallery focusing on animal
imagery. New and established artists.
Regular exhibitions

*LONDON GLASSBLOWING WORK-SHOP
7 The Leather Market
Weston Street
London Bridge
SE1 3ER
Tel: 0171 403 2800
Fax: 0171 403 7778
Contact: Peter Layton
Established: 1976
Open: 10am to 6pm Monday to Friday
(weekends by appointment)
Craftworkers on site: 6-7
Crafts: Freeblown coloured studio glass
Prices: £3 to £300+. American Express/
Visa/Mastercard/Access
Note: Demonstrations held. Open
weekends. Internationally renowned for

variety of form, colour and texture.
Disabled access to workshop (lift), five
steps to gallery

JAMES TAYLOR & SON
4 Paddington Street
Marylebone
W1M 3LA
Tel: 0171 935 4149
Fax: 0171 486 4212
Contact: Peter Schweiger
Established: 1857
Open: Monday to Friday 9am to 5.30pm
and Saturday 10am to 1pm
Craftworkers on site: 12
Crafts: Footwear
Prices: £50 for ready made sandals and
bespoke from £795 + VAT. Most major
credit cards accepted
Note: Tours of the workrooms can be
arranged. Disabled access. Madam
Tussauds nearby

ELECTRUM
21 South Molton Street
Mayfair
W1Y 1DD
Tel: 0171 629 6325
Contact: Barbara Cartlidge
Established: 1971
Open: Monday to Friday 10am to 6pm.
Saturday 10am to 1pm
Stock: Contemporary jewellery by
artists from Britain and abroad includ-
ing Ramshaw, Flöckinger, Preston,
Fisch, Viegeland, Illsley, Itoh, Betts and
Torun
Prices: From under £ 30 to over £ 2000
Major credit cards accepted

*GALERIE BESSON
15 Royal Arcade
28 Old Bond Street
Mayfair
W1X 3HB
Tel: 0171 491 1706
Fax: 0171 495 3203
Contact: A. Besson
Stock: Studio Ceramics. Permanent stock
of classic potters including Coper, Rie,
Leach etc. Of a quality and variety which
would be of interest to collectors
Open: Monday 1pm to 5.30pm. Tuesday
to Friday 10am to 5.30pm. Saturday
10am to 12.30pm during exhibitions
Prices: From under £50 to over £5000
Credit cards accepted

EMMA HICKMOTT CERAMICS
18 Portobello Green Arcade
281 Portobello Road
North Kensington
W10 5TD
Tel: 0181 968 5800
Contact: Emma Hickmott
Established: 1990
Open: Tuesday to Saturday 10am to 5pm
Crafts: Handpainted and transfer decorated bone china gift and tableware, produced on the premises
Prices: £4.50 to £45. American Express/Visa/MasterCard/Eurocard

*CECILIA COLMAN GALLERY
67 St. John's Wood
High Street
St. John's Wood
NW8 7NL
Tel: 0171 722 0686
Contact: Cecilia Colman
Established: 1979
Open: Monday to Friday 10am to 5.30pm. Saturday 2pm to 5pm. Saturdays in December until Christmas 10am to 5.30pm
Stock: Wide selection of British ceramics, glass and jewellery. Unusual mirrors and carved wood.
Prices: £30 to £1,500
Credit cards
Note: Regular exhibitions

CONTEMPORARY APPLIED ARTS
2 Percy Street
Soho
W1P 9FA
Tel: 0171 436 2344
Contact: Yvonne Kulagowski
Established: 1948
Open: Monday to Saturday 10.30am to 5.30pm
Stock: Ceramics, glass, wood, metal, jewellery, textiles, furniture and wood
Prices: £20 to £20,000
Note: Membership association with regular exhibitions and shop. Commissioning service available

CONTEMPORARY CERAMICS
7 Marshall Street
Soho
W1V 1LP
Tel: 0171 437 7605
Fax: 0171 287 9954
Contact: Marta Donaghey
Established: 1960
Open: 10am to 5.30pm Monday to Saturday. 10am to 7pm Thursday
Stock: Domestic pottery, studio ceramics, jewellery. Ceramics made by members of the Co-operative Craft Potters Association
Prices: £5 to £1,000. Visa/Access/Diners/Amex/Switch
Note: Regular exhibitions held of new work. Large stock of books on ceramics

HERITAGE CERAMICS
Unit B18
Charles House
Bridge Road
Southall
Middlesex
UB2 4BD
Tel: 0181 843 9281
Fax: 0181 813 8387
Contact: Tony Ogogo
Established: 1984
Craftworkers on site: 2
Open: 10am to 6pm weekdays
Crafts: African pottery and handcrafted African design-inspired tableware
Prices: £5 to £200. Credit cards not accepted
Note: Demonstrations of African traditional pottery techniques. Children and people with special needs are welcome

THE GLASS STUDIO
208 Second Floor
Oxo Tower Wharf
Bargehouse Street
Southwark
SE1 9PH
Tel: 0171 633 9267
Fax: 0181 948 2412
Contact: Devi Khakhria
Established: 1996
Open: 11am to 6pm Tuesday to Sunday
Craftworkers on site: 1
Crafts: Handmade glass plates, china tableware, handmade cards
Prices: £1.40 to £135. Credit cards not accepted
Note: Individual designs by Devi Khakhria on functional objects using specialist technology and bright uplifting colours. Situated in new development around listed Art Deco OXO Tower. Other craftspersons on site. Disabled access

BANKSIDE GALLERY
48 Hopton Street
Southwark SE1 9JH
Tel: 0171 928 7521
Fax: 0171 928 2820
Contact: Richard Russell
Established: 1985 (Gallery). Societies:
RWS 1804, RE 1880
Open: During exhibitions: Tuesday 10am
to 8pm, Wednesday to Friday 10am to
5pm. Sunday 1pm to 5pm. Saturday 1pm
to 5pm. Outside exhibitions: bookshop
only 10am to 4pm Tuesday to Friday
Stock: Watercolours and original artists'
prints from members of the Royal
Watercolour Society and Royal Society of
Painter-Printmakers. Occasionally
ceramics
Prices: £25 to £2,000. Visa/Delta/
Eurocard/MasterCard/Switch
Note: Exhibitions, demonstrations, talks,
tours, open days etc. Bookshop and light
refreshments

UNION HALL WORKSHOPS
27-29 Union Street
Southwark SE1 1SD
Tel: 0171 357 7085
Fax: 0171 357 7085
Contact: A. Cardew
Established: 1993
Open: 9am to 6pm
Craftpersons on site: 3
Crafts: Furniture and other wood related
products
Prices: £15 to £3,000. Credit cards not
accepted
Note: Furniture mostly made to
commission with smaller items of
woodware for sale

VIVIENNE LEGG CERAMICS
16 Gabriel's Wharf
56 Upper Ground
Southwark
SE1 9PP
Tel: 0171 401 2240
Contact: Vivenne Legg
Established: 1989
Open: Tuesday to Sunday 10am to 6pm
Craftworkers on site: 1
Crafts: Pottery
Prices: £12 to £130. Visa/Access/
MasterCard/Amex accepted
Note: Courtyard with other craft shops,
restaurants, and sandwich bars. Disabled
access to courtyard

CRAFTS COUNCIL SHOP
Victoria and Albert Museum
South Kensington
SW7 2RL
Tel: 0171 589 5070
Contact: Kathleen Salter
Established: 1974
Open: Tuesday to Sunday 10am to
5.30pm. Monday 12 noon to 5.30pm
Stock: Ceramics, glass, jewellery, textiles,
wood, metalwork
Prices: £15 to £1,500. Amex/Visa/
Mastercard
Note: Victoria & Albert Museum, Cafe,
changing exhibitions in shop

LIVING GLASS STUDIO
67 Fulwell Road
Teddington
Middlesex TW11 0RM
Tel: 0181 977 5277
Contact: Barry Kendall
Established: 1978
Open: 10am to 6pm Tuesday, Thursday,
Friday and Saturday
Craftworkers on site: 1
Crafts: Stained glass, panels, lamps and
3D pieces
Prices: £8 to £500. Credit cards not
accepted

EXTRAORDINARY DESIGN
Top Floor
Tudorleaf Business Centre
2-4 Fountayne Road
Tottenham
N15 4QL
Tel: 0181 808 3969
Fax: 0181 808 3969
Contact: Dan Maier
Established: 1992
Open: By appointment only – always ring
first
Craftworkers on site: 3
Crafts: Knitted chandeliers and other
pendant light shades, tablelamps and
textile installations, knitwear, jewellery
and hats
Prices: £20 to £500. Credit cards not
accepted

WHITECHAPEL POTTERY
2A East Mount Street
(opposite Whitechapel Tube Station)
Whitechapel
E1 1BA
Tel: 0171 377 2171

Contact: Frances Bingham
Open: 11am to 6pm Tuesday to Saturday
Craftworkers on site: 1
Crafts: One-off pieces, both functional
and sculptural, in white stoneware by Liz
Mathews
Prices: £12 to £500. Access/Visa
accepted
Note: Phone for details of studio open
days, private views, invitation events.
Disabled access

stop press

MAZE HILL POTTERY
The Old Ticket Office
Woodlands Park Road
Greenwich
SE10 9XE
Tel: 0181 298 0048
Contact: Lisa Hammond/John Dawson
Established: 1994
Open: 10am to 5pm
Craftworkers on site: 3
Crafts: Excellent quality domestic
functional and one off soda glaze
stoneware and porcelain
Prices: £3.50 to £250 , Most major cards
accepted
Note: There is disabled access and
pottery is near the main line Station from
Charing Cross

In addition to the above there are a number
of workshop centres in London where groups
of craftworkers, of varying numbers, have
studios. In general these do not operate as
retail units but we include the following
contact details

401½ WORKSHOPS
401½ Wandsworth Road
SW8

CLERKENWELL WORKSHOPS
27 Clerkenwell Close
EC1R 0AT

CLOCKWORK STUDIOS
38 Southwell Road
SE5 5PG

COCKPIT STUDIOS
Cockpit Yard
Northington Street
WC1N 2NP

ILIFFE YARD SUDIOS
Crampton Street
SE17 3QA

KINGSGATE WORKSHOPS
110-116 Kingsgate Road
NW6 2JG

BROOKFIELD CRAFT CENTRE
333 Crumlin Road
Belfast
BT14 7EA
Tel: 01232 745241
Fax: 01232 746431
Open: 9am to 5pm Monday to Saturday
Crafts: Jewellery, knitwear, ornate
sculpture, hurley sticks, traditional Irish
bodhrans

COPPER MOON
The Spires, Howard Street
Belfast
Tel: 01232 235325
Open: 9am to 5.30pm Monday to
Saturday. 9pm Thursday
Stock: Contemporary handcrafts and
jewellery. Wooden games, ornately styled
pottery, hand-printed silk scarves,
candlestick holders, decorative mirrors,
unique paintings and cards

CRAFTWORKS LTD
Unit 8
Bedford House, Bedford Street
Belfast
BT2 7FD
Tel: 01232 244465
Open: 9.30am to 5.30pm Monday to
Saturday. 8.30pm Thursday
Stock: Ceramics, jewellery, hand-painted
silk scarves, handwoven waistcoats and
hats, glass, wood, musical instruments,
handwoven baskets, clocks, hand-tied
fishing flies

THE STEENSONS
Bedford House, Bedford Street
Belfast
Tel: 01232 248269
Open: 10 am to 5.30 pm Monday to
Saturday
Crafts: A fine and wide selection of
contemporary jewellery in precious metals
Notes: Thought to be one of Ulster's
leading galleries for this craft type

ULSTER WEAVERS
44 Montgomery Road
Castlereagh
Belfast
Tel: 01232 404236
Open: 9.30am to 5pm Monday to Saturday
Stock: Linen tablecloths, napkins, sheets,
tea towels, kitchen co-ordinates. Local
crafts include pottery, glassware, linen
lampshades

BALLYGALLEY CRAFT SHOP
273 Coast Road, Ballygalley
County Antrim
Tel: 01574 583640
Open: Open daily 11.30am to 5.30pm
Easter to October. Winter 11.30am to
5.30pm. Saturday from 1.30pm.
Stock: Traditional local handcrafts. Toys,
knitwear, woodturning stools, patchwork
cushions, dried flower arrangements,
jewellery, prints of the Glens of Antrim

CRANE BAG GALLERY
15 Broughshane Street, Ballymena
County Antrim
Tel: 01266 40569
Open: 9.30am to 5.30pm Monday to
Saturday
Stock: Local ceramics including Forge
Pottery kitchen and domestic stoneware,
designer jewellery, original watercolours
set in brooches and pendants

FORGE POTTERY
18 Milltown Road
Antrim
County Antrim
BT41 4NW
Tel: 01849 465349
Contact: Linda Murphy
Open: 9am to 5pm Monday to Friday but
please telephone in advance
Crafts: Kitchen and domestic stoneware
in 'Apple' and 'Snakeskin' designs, hand-
built and painted in warm colours
Note: Leave M2 at Ballymena exit, take
A6 from roundabout. Road 200 yards on
right, pottery after one-and-a-half miles

PATTERSON'S SPADE MILL
Antrim Road
Templepatrick
County Antrim
Tel: 01849 433619
Open: April, May and September
weekends 2pm to 6pm. High summer
daily 2pm to 6pm (excluding Tuesdays)
Crafts: Turf and garden spades
Prices: approximately £30
Notes: Demonstrations held and special
spades made to order

SEAFRONT EXHIBITION CENTRE
Mary Street
Ballycastle
County Antrim
BT54
Tel: 01265 762225
Open: Daily June to September and
weekends Easter to September
Stock/Crafts: Designer screen painting,
handpainted screens, patchwork quilts
and waistcoats, lace, decorated stones,
carved wooden lampstands, wrought
ironwork

THE STEENSONS
64 Toberwine Street
Glenarm
County Antrim
Tel: 01574 841445
Open: 9.15am to 5.15pm Monday to
Friday plus weekends in summer
Crafts: A fine selection of mainly silver
contemporary jewellery. Also at Bedford
House, Belfast -see previous page

BALLYDOUGAN POTTERY
133 Plantation Road
Portadown
Craigavon
County Armagh
BT63 5NN
Tel: 01762 342201
Contact: Sean O'Dowd
Open: 9am to 6pm Monday to Saturday
Crafts: Domestic ware – hand-thrown
pottery cups, saucers, plates, dishes and
bowls inspired by the flora and fauna of
Ireland
Notes: Located in 300 year-old farm-
house. Leave M1 at Moira exit, follow A3
to Lurgan. At first set of traffic lights turn
left to roundabout, take road for Gilford,
Pottery 3 miles on left

BEECHMOUNT IRONCRAFT
23 Moygannon Lane
Donaghcloney
County Armagh
BT66 7ND
Tel: 01762 881162
Contact: Mr. P. Brooker
Established: 1979
Craftworkers on site: Yes
Open: 8.30am to 5.30pm Monday to
Friday. 8.30am to 1pm Saturday
Stock/Crafts: Hand made articles for sale
from the showroom

JUDITH LOCKHART POTTERY & PORCELAIN
The Beeches
Derry Willigan
211 Armagh Road, Newry
County Armagh
BT35 6NW
Tel: 01693 821362
Contact: Judith Lockhart
Established: 1980
Craftworkers on site: Yes
Open: Most days but please phone first
Stock/Crafts: Porcelain and terracotta.
Limited edition pieces. Specialist gift
items
Notes: Commissions undertaken

PALACE STABLES HERITAGE CENTRE
Palace Demesne
Armagh
County Armagh
Tel: 01861 529629
Open: Daily
Stock: Locally produced contemporary
crafts–pottery, knitwear, lace, wood craft,
jewellery, books of local interest

RATHBANNA COPPER WORKS
23 Derrinraw Road
Portadown
Craigavon
County Armagh
BT62 1UX
Tel: 01762 851281
Contact: Eddie and Paul O'Neill
Established: 1895
Stock/Crafts: Copper wall murals,
pictures and découpage

TULLYLISH POTTERY
59a Banbridge Road
Tullylish, Gilford
County Armagh
BT63 6DL
Tel: 01762 831765
Contact: Raymond McMillen Boyd
Open: 10am to 6pm Monday to Saturday
Crafts: Hand-decorated stoneware made
in adjoining pottery. Functional and
decorative domestic ware in earthy
colours

BURREN HERITAGE CENTRE
15 Bridge Road
Burren
Warrenpoint
County Down

Tel: 01693 773378
Open: Tuesday to Sunday, July and August
Crafts: Local crafts and souvenirs, hand painted scarves, jewellery, crochet and pottery.

CELTIC CRAFTS GALLERY
45 Dromara Road
Dundrum
Newcastle
County Down
Tel: 01396 751327
Contact: Mary Doran
Open: 10am to 5pm, Monday to Saturday, all year
Stock: Jewellery in silver and gold
Note: Beautiful setting about a mile from Dundrum village with views to the mountains of Mourne. Children's play area and coffee shop. Workshop demonstrations possible.

CONEY ISLAND DESIGN
41 Killough Road
Coney Island
Ardglass
County Down
BT30 7UG
Tel: 01396 841673
Contact: Berni Sutton
Established: 1981
Open: Please 'phone first
Stock: Hand painted tiles and ceramic sculptures

COWDY CRAFTS
20 Main Street
Hillsborough
County Down
BT26 6AE
Tel: 01846 682455
Contact: Patrick Cowdy
Established: 1969
Open: 9.30am to 5.30pm Monday to Saturday all year.
Crafts: Modern pewter, jewellery, glass, hand made pottery, woven textiles
Prices: Reasonably priced gifts
Visa/Access/Switch accepted

CRAIGMOUNT
Inch Lodge
75 Belfast Road
Downpatrick
County Down
BT30 9AY

Tel: 01396 612542
Contact: Patricia Cross
Stock/Crafts: Celtic jewellery
Access/Visa accepted

DISCOVERY GLASS
9 High Street
Comber
County Down
Tel: 01247 870181
Open: Monday to Saturday normal business hours
Crafts: Jewellery, light catchers, boxes, lamps and windows in handcrafted stained glass

EDEN POTTERY
39a South Street
Newtonards
County Down
BT23 4JT
Tel: 01247 820372
Contact: Heather Anderson
Open: 9.30am to 5.30pm, Monday to Saturday
Crafts: Hand thrown and decorated tableware – plates, bowls etc as well as garden pots

FERGUSON LINEN CENTRE
54 Scarva Road
Banbridge
County Down
BT32 3AU
Tel: 01820 623491
Contact: Siobhan Lennon
Established: 1854
Open: Monday to Friday usual business hours. Factory tours at 11am and 3pm. Friday only 11am. £2 per person. Groups by arrangement
Crafts: World's only manufacturer of double damask linen. All types of linen items available in the shop

KILLYLEAGH YARNS
Catherine Street
Killyleagh
County Down
BT30
Tel: 01396 828204
Stock/Crafts: Flax spinning and winding to make yarns which are used in table cloths, napkins, tray cloths, tea towels, aprons etc.
Notes: Factory tour including video can be made by prior arrangement

LOCH RURAY HOUSE
8 Main Street
Dundrum, Newcastle
County Down
Tel: 01396 751544
Open: 11am to 5.30pm, Tuesday to
Sunday, March to December but
available other times by appointment
Crafts/Stock: Ceramics, jewellery,
paintings, turned wood as well as linen
items from the workshop

MONLOUGH WORKSHOP LTD
19 Monlough Road West
Ballygowan, Newtownards
County Down
BT23 6ND
Tel: 01232 813451
Contact: Andrew Klimacki
Established: 1985
Open: 8am to 5pm Monday to Friday
Crafts: Furniture designed and made
Prices: On application. Credit cards not
accepted

PARROT LODGE POTTERY
131 Ballyward Road
Ballyward, Castlewellan
County Down
BT31 9PS
Tel: 01820 650314
Contact: Karen Noble
Craftworkers on site: Yes
Open: 10am to 5pm, Monday to Saturday
Stock/Crafts: Hand thrown and decorated
stoneware pottery

PINEWOOD POTTERY
88 Cootehall Road
Crawfordsburn
Bangor
County Down
BT19 1UW
Tel: 01247 852264
Contact: Norbert Kus
Established: 1990
Open: 9am to 5pm Monday to Friday
Stock/Crafts: Tableware and children's
tableware
Prices: 50p to £35. Credit cards accepted

RED COTTAGE CRAFTS
1 Rawdon Court
Main Street
Moira
County Down
BT67 0LQ

Tel: 01846 619172
Contact: Jean Neill
Open: Monday to Friday 9.30am to
5.30pm
Stock/Crafts: Patchwork quilts, cushions,
American fabrics and tapestries
Prices: Various. Credit cards accepted

THE BAY TREE
Audley Court
118 High Street
Holywood
County Down
BT18
Tel: 01232 426414
Open: 10am to 4.30pm Monday to
Saturday
Stock: Mainly ceramics gallery featuring
stock from all over Ireland in a wide
variety of styles, patterns and colours
Notes: Coffee shop adjacent

TURNIP HOUSE KNITWEAR
Causeway Road
Newcastle
County Down
BT33 0DL
Tel: 013967 26754
Contact: Elaine and John McCombe
Established: 1987
Open: 10am to 6pm, Monday to Friday
and 2pm to 6pm on Saturday.
Craftworkers on site: Yes
Stock/Crafts: Wide range of colourful
knitwear with celtic motifs
Prices: £40 to £100. Access/Visa/
MasterCard accepted
Notes: No disabled access

WILD FLOWER GALLERY
31 Mallymoney Street
Ballymena
County Down
BT43
Tel: 01266 46516
Open: 9.30 am to 5.30 pm, Monday to
Saturday, closed Wednesday
Stock: Hand painted wooden utensils,
ceramics and jewellery

ANNE MCNULTY STONEWARE
Unit 2, Enniskillen Enterprise Ctre
Down Street
Enniskillen
County Fermanagh
BT74 7DU
Contact: Anne McNulty

Open: 9.30am to 5.30 pm, Monday to Saturday
Crafts: White stoneware tableware decorated with brightly coloured brush work
Notes: Coffee shop

BELLEEK CRAFT VILLAGE
Main Street, Belleek
County Fermanagh
BT93 3FY
Tel: 01365 658631
Fax : 01365 658631
Contact: Brian Flannigan
Established: 1991
Craft workers on site: 3 at time of going to press but will rise to 7
Open: 10am to 5pm Monday to Saturday
Stock/Crafts: Glass (being a full range of table and giftware), leather and decorative steel candle sticks
Prices: From under £10 to over £650
Most major credit cards accepted
Notes: There are a number of craft producers grouped on this site. A 15 minute tour of the glass workshop (Fermanagh Crystal) is also possible

BELLEEK POTTERY LTD
Belleek, Enniskillen
County Fermanagh
BT74 7AA
Open: Shop open daily from March to September but only weekdays from October to February
Stock/Crafts: A wide range of parian type ware for which this Pottery is famous
Notes: Restaurant, museum, video presentation and tours of the studio

BOSTON QUAY
The Buttermarket
Down Street
Enniskillen
County Fermanagh
Tel: 01365 324499
Contact: Cathy Brittain
Established: 1991
Open: 9.30am to 5.30pm, Monday to Saturday
Craftworkers on site: 15
Stock/Crafts: Wide range of Irish crafts including jewellery, handpainted and woven textiles, leather work, furniture, glass, toys and knitwear
Prices : From under £5 to over £200.
Major credit cards accepted
Notes: Mail order service also available

DOUNESYDE CRAFT SHOP
Sligo Road, Enniskillen
County Fermanagh
BT74 5QN
Contact: Rosalie Little
Established: 1987
Open: Monday half day and all day Tuesday to Saturday

FERMANAGH COTTAGE INDUSTRIES
14 East Bridge Street, Enniskillen
County Fermanagh
BT74
Tel: 01365 322260
Open: 9.30 am to 5.30 pm, Monday to Saturday
Crafts: Traditional handmade Irish crafts. Irish blackthorn walking sticks, celtic jewellery, hand cut glass, cross stitch items and leather work

GILMARTINS THE CRAFT SHOP
Main Street
Belleek
Enniskillen
County Fermanagh
BT93 3FY
Tel: 013656 58371
Open: 9am to 6pm Monday to Saturday plus Sundays in summer
Crafts: Designer crafts, hand tied fishing flies framed or for use as brooches, Beleek china and general craft gift items

THE YELLOW DUCK
The Diamond, Enniskillen
County Fermanagh
BT74
Tel: 01365 326987
Open: Monday to Saturday, 9am to 6pm
Crafts: Mainly ceramics showing both domestic and figurative work

MCCLUSKEY POTTERY
11 Gortgarn Road, Limavady
County Londonderry
BT49 0QW
Tel: 01504 764579
Contact: Sean McCluskey
Open: 9am to 5pm Monday to Friday Also open at weekends during summer but please phone for times
Stock/Crafts: Hand made domestic ware, lamp bases and giftware decorated often with floral motifs
Notes: 3 miles from Limavady on the A37

ROWEN ROCKING HORSES
56 Balloughry Road
Londonderry
County Londonderry
BT48 9XL
Tel: 01504 264715
Contact: Wendy Mackey
Note: Disabled access to workshop and
house

THE CRAFT CENTRE
25 The Village
Shipquay Street
Derry
County Londonderry
BT48
Tel: 01504 261876
Open: 10am to 5.30pm, Monday to
Saturday
Stock: Contemporary studio crafts,
particularly ceramics, celtic design
pieces, batik and cards

AN CREAGAN VISITOR CENTRE
Creggan
County Tyrone
Tel: 01662 761112
Contact: The Manager
Stock: Extensive range of pottery, local
paintings, jewellery, celtic items,
handcarved sculptures in local bog wood
and various souvenir gift items

MILL POTTERY
59a Brook Road
Dunamanagh
County Tyrone
BT82 0RX
Contact: Tom Agnew
Open: 9am to 6.30pm, Monday to Friday
Stock/Crafts: Hand thrown domestic
stoneware
Notes: Building is an old flax mill

SUITOR GALLERY
Ballygawley Roundabout
Ballygawley
County Tyrone
Tel: 01662 568653
Contact: Beryl Suitor
Craftworker on site: 1
Open: 10am -5.30pm, Tuesday to
Saturday
Stock/Crafts: Ceramics, celtic jewellery,
rugs, handmade candles, prints of
Northern Ireland views.
Notes: Tea shop with wheelchair access

TYRONE CRYSTAL CRAFT CENTRE
Killybrackey
Coalisland Road
Dungannon
County Tyrone
Tel: 01868 725335
Open: 9.30am to 3.30pm, Monday to
Saturday. Winter weekdays only
Stock/Crafts: Range of cut glass
Notes: Tours of the works are made every
30 minutes during the above hours when
the entire process of blowing and
finishing glass can be seen
Restaurant

SPEYSIDE POTTERY
Ballindalloch
Aberlour
Banffshire AB37 9BJ
Tel: 01807 500338
Contact: Ann or Thomas Gough
Open: 10am to 5pm April to September
Craftworkers on site: 2
Crafts: Wide variety of handthrown
functional woodfired stoneware
Prices: From under £5.00 to over £70
Credit cards not accepted.

WALTER AWLSON
31 Ludgate
Alloa
Clackmannanshire FK10 1DP
Tel: 01259 214304
Contact: Walter Awlson
Established: 1991
Open: 8am to 5pm Monday to Friday
Craftworkers on site: 1
Crafts: Figurative ceramic sculptures for
the home or garden, slip-cast in limited
editions
Prices: £80 to £500

A. M. WOODWARE
2 Stuart Street
Ardersier
Inverness-shire IV1 2QL
Contact: Alisdair McKay
Established: 1984
Open: Most working hours and by
arrangement
Craftworkers on site: 1
Crafts: Furniture and some woodware
Prices: £5 to £5,000. Credit cards not
accepted
Note: Most work by commission

PAINTINGS & PULLOVERS
Crofts Cottage, Glenmuick
Ballater
Aberdeenshire AB35 5SU
Contact: H. Butterworth
Established: 1977
Open: Daily 10am to 5pm
Craftworkers on site: 2
Crafts: Paintings, prints, cards and
knitwear
Prices: £1 to £10,000. Access/Visa/
Mastercard/Eurocheque
Note: Peafowl, poultry and ducks
roaming free around natural gardens

A'ANSIDE STUDIOS
5 Main Street
Tomintoul
Ballindalloch
Banffshire
AB37 9EX
Contact: Barry Horning
Tel: 01807 580430
Established: 1996
Open: 9am to 6pm Monday to Saturday
10am to 6pm Sunday
Craftworkers on site: 2
Crafts/stock: Selected craftwork from
N.E. Scotland and Highlands
Prices: 99p to £500+
Note: Visit stained glass workshop.
Courses held

MADE IN SCOTLAND
The Craft Centre
Station Road
Beauly
Inverness-shire
IV4 7EH
Tel: 01463 782821
Fax: 01463 782409
Established: 1991
Open: Summer: 10am to 5.30pm
Monday to Saturday, 12 noon to 5pm
Sunday. Winter: 10am to 5pm Monday to
Saturday, closed Sunday
Stock: Scottish crafts, knitwear, textiles
Prices: 30p to £200. Visa/Access/
American Express

BROUGHTON GALLERY
Broughton Place, Broughton
Biggar
Lanarkshire ML12 6HJ
Tel: 01899 830234
Contact: Graham Buchanan-Dunlop
Established: 1976
Open: Daily 10.30 am to 6pm closed
Wednesday. Closed from Christmas to
late March and part of November. Please
phone for exact dates before visiting at
this time of year.
Stock: High quality domestic pottery,
studio ceramics, glass, wood, metal,
jewellery, textiles, furniture and toys as
well as original prints and paintings
Note: Building remarkable castle type
appearance designed by Sir Basil Spence
Prices: From under £10 to over £2000.
Major credit cards accepted

BUCHLYVIE POTTERY SHOP
Main Street
Buchlyvie
Stirlingshire
FK8 3LP
Tel: 01360 850405
Contact: Alison Borthwick
Established: 1983
Open: 9.30am to 5.30pm Monday to
Saturday. 12.30pm to 5pm Sunday
Stock: Fine, hand-painted, slip-cast
porcelain including tableware, giftware,
lamps
Prices: £5 to £75
Note: Disabled access

R. GRANT LOGAN
Glenramskill
Campbeltown
Argyll PA28 6RD
Tel. & Fax: 01586 553588
Contact: Grant Logan
Established: 1978
Open: Monday to Friday 9am to 1pm and
2pm to 5pm. Saturday 9am to 12.30pm
only
Craftworkers on site: 2
Crafts: Precious gold jewellery incorpo-
rating gem stones
Prices: £30 to £2,000. Visa/Mastercard/
Access
Note: Goldsmith can be seen at work

ANNE HUGHES POTTERY
Auchreoch
Balmaclellan
Castle Douglas
Kirkcudbrightshire
DG7 3QB
Contact: Anne Hughes
Tel: 01644 420205
Open: 10am to 6pm from Easter to
September
Craftworkers on site: 1
Crafts: The wide range of work is
unusual, colourful and varied, with
pierced flower plates a speciality. Some
necklaces available
Prices: From under £3 to over £70
Note: Commissions undertaken

ANNIE HORSLEY POTTERY
Laurieston Hall
Castle Douglas
Kirkcudbrightshire
DG7 2NB
Tel: 01644 450657

Open: Tuesday & Wednesday 10am to 5pm
Craftworkers on site: 1
Crafts: Domestic and decorative
earthenware
Prices: £2.00 to £100

GALLOWAY FOOTWEAR
The Clog and Shoe Factory
Balmaclellan
Castle Douglas
Kirkcudbrightshire DG7 3QE
Tel: 01644 420465
Fax: 01644 420777
Contact: Godfrey Smith
Established: 1979
Open: Monday to Friday 9am to 5pm
Easter to October. Off-season and
weekends please phone first
Craftworkers on site: 2/3
Crafts: Clogs, boots, shoes and sandals.
Each pair of footwear can be individually
made for you
Prices: £40 to £200
Note: Commissions welcome. Disabled
access

NORTH GLEN GALLERY
Palnackie
Castle Douglas
Kirkcudbrightshire DG7 1PN
Tel: 01556 600200
Contact: Ed and Tom Iglehart
Open: 10am to 6pm most days in summer
but phone call advisable in off season
Crafts: Decorative and functional blown
glass and metalwork. Tree houses,
teepees and experimental structures
Prices: £5 to £5000
Note: Located on A711 south west of
Dalbeattie

TREVOR LEAT BASKETS
The Hill, Balmaclellan
Castle Douglas
Kirkcudbrightshire DG7 3PW
Tel: 01644 420682
Contact: Trevor Leat
Open: Most days but please phone first
Craftworkers on site: 1
Crafts: Traditional & contemporary baskets
Prices: £10 to £100

MARI DONALD KNITWEAR
Pudding Lane
Comrie
near Crieff
Perthshire PH6 2DS

Contact: Roddy or Mari Donald
Established: 1983
Open: Monday to Friday and Saturday mornings. Hours vary according to time of year
Craftworkers on site: 2
Crafts: Children's and adults' knitwear, paintings, woodwork, papier mâché
Prices: £1 to £75. Visa/Mastercard

THE GREEN HALL GALLERY
2 Victoria Street
Craigellachie
Banffshire AB38 9SR
Tel and Fax : 01340 871010
Contact: Stewart Johnston/Maggie Riegler
Established: 1992
Open: Monday to Saturday 9am to 1pm and 2pm to 5.30pm. Sundays 2pm to 5.30pm
Craftworkers on site: 2
Crafts: Handmade cards on handmade paper, prints on handmade paper, oils and watercolours
Prices: £1 to £350. Visa/Mastercard/ Eurocard
Note: Our gallery is an old gospel hall with lovely views over the Spey – steps from garden to Speysial Way. B & B next door to gallery

THE STREATHEARN GALLERY & POTTERY
32 West High Street
Crieff
Perthshire PH7 4DL
Tel: 01764 656100
Fax: 01764 654126
Website: www.strathearn-gallery. com
E-mail: info@strathearn-gallery.com
Contact: Edith and Owen Maguire
Established: 1994
Open: 10am to 5pm all year
Stock: High quality contemporary domestic pottery, studio ceramics, glass, wood, metal and jewellery
Prices: £10 to £1,000. All major credit cards accepted
Note: Regular exhibitions held all year

BARNBARROCH POTTERY
Kippford
Dalbeattie
Kirkcudbrightshire DG5 4LE
Tel: 01556 620695
Contact: Christine or Rodger Smith
Established: 1978

Open: 9.30am to 5.30pm Monday to Saturday
Craftworkers on site: 3
Crafts: Decorated earthenware. Hand thrown terracotta garden pots
Prices: £1.50 to £300. Credit cards not accepted
Note: Situated in a scenic area half a mile from the sea

HIGHLAND TWEEDS
Highland Tweed Manufactory
86 High Street
Dingwall
Ross-shire IV15 9TF
Contact: M. Fraser
Established: 1973
Open: 9am to 6.30pm Monday to Friday. Saturday 9am to 2pm
Craftworkers on site: 3
Crafts: Tweeds, tartans, travel rugs, scarves, stoles, headsquares
Prices: £2 to £30. Credit cards not accepted
Note: Demonstrations held

ICEBERG GLASSBLOWING STUDIO
Victoria Buildings
Drumnadrochit
Inverness-shire IV3 6TU
Contact: Julie Snowdon
Established: 1984
Open: Summer 9am to 9pm 7 days. Winter 10am to 5pm 6 days
Craftworkers on site: 2
Crafts: Lamp worked glassblowing, jewellery, etc.
Prices: £1 to £15. Visa/MasterCard/Delta
Note: Loch Ness exhibition nearby

TOUCH WOOD
Gordon House
Church Street
Dufftown
Banffshire AB55 4AR
Tel: 01340 820996
Contact: Stephen Gould
Established: 1995
Open: 9am to 6pm or by appointment
Craftworkers on site: 1
Crafts: Celtic bowls, platters, plaques, lamps, celtic plaques
Prices: £15 to £100. Credit cards not accepted
Note: Native woods used: ash, oak, elm, sycamore, beech, chestnut. Commissions undertaken

DAVID GULLAND GLASS
The Old Coach House, 5 Rotchell Road
Dumfries
DG2 7SP
Tel: 01387 251492
Contact: David Gulland
Established: 1978
Open: 10am to 12.30pm and 2.30pm to 5pm Tuesday to Friday
Craftworkers on site: 1
Crafts: Copper wheel engraving on crystal, dry sand glass and holloware including commissions for public awards
Prices: £15 to over £1000
Note: Located opposite camera obscura and museum. Disabled access

HARMANS
72 Brooms Road
Dumfries
DG1 2AH
Tel: 01387 252971
Established: 1987
Open: 10am to 5.30pm Monday to Saturday
Crafts: Handmade (knotted) rugs
Prices: £25 to £3,000+
Note: Hand-knotted rug making and appreciation classes throughout the year

LAURISTON HOUSE POTTERY
Wallaceton
Dunscore
Dumfries
DG2 OTG
Tel: 01387 820472
Contact: Jason Shackleton
Open: 9am to 5pm daily all year round
Stock: Domestic earthenware, one-off pieces and tile panels. Decorated with slip, tin glaze and all wood fired.
Prices: £5.00 upwards

EDUARDO ALESSANDRO GALLERY
30 Gray Street, Broughty Ferry
Dundee
Tayside DD5 2BJ
Tel: 01382 737011
Fax: 01382 737012
Contact: Mr. S. Paladini
Established: 1977
Open: 9.30am to 5pm Monday to Saturday. 9.30am to 1pm Wednesdays in winter. Closed Sunday
Stock: Pottery, ceramics, sculpture, jewellery, floral art etc. Paintings and prints by Scottish artists and craftsmen

Speciality: Paintings, prints and pottery by Scottish artists
Prices: Under £10 to £3,000. Access/ Visa/Switch
Note: One of the major art galleries on the east coast of Scotland

THE WEE CARD SHOP GALLERY
2 Campfield Square, Barnhill
Dundee
Tayside DD5 2PU
Tel: 01382 736384
Contact: Mrs Maureen Nairn
Established: 1983
Open: 9am to 5.30pm Monday to Saturday
Stock: Ceramics, wooden ornaments, porcelain, velvet shoulder bags, earrings, posters and prints
Speciality: Greetings cards
Prices: £1 to £35. Access/Visa/Mastercard
Note: Many gifts are designed and crafted in Scotland

CERAMICS WORKSHOP
38 Merchiston Avenue
Edinburgh EH10 4NZ
Tel: 0131 228 3016
Contact: Liz Rosenthal
Established: 1984
Open: 10am to 5.30pm Monday to Friday. 11.30am to 5.30pm Saturday
Craftworkers on site: 3
Crafts: Ceramics, plaster work, jewellery, cards, pictures
Prices: £1.50 to £150. Credit cards not accepted
Note: Majority of work made by Liz Rosenthal, Lucy Iredale, Helen Spalding

CHARROSA
Unit 5, Castlebrae Business Centre
Peffer Place
Edinburgh
EH6 8SR
Contact: Charlotte Macdonald
Established: 1994
Open: 10am to 6pm Monday to Friday and by appointment
Craftworkers on site: 1
Crafts: Hand painted silk ties, waistcoats, bow ties, braces, pictures/wall hangings (commissioned), scarves and spectacle cases
Prices: £7 to £90
Note: Demonstrations arranged. Items can be personalised or specially designed

RAGAMUFFIN
276 Canongate
Royal Mile
Edinburgh
EH8 8AA
Tel: 0131 557 6007
Established: 1975
Open: 10am to 6pm or later, seven days a week
Stock: Designer knitwear from Britain and Ireland, Fair Isles, chunkys, handknits and Shetlands. Our own designs in tweeds and silks. Pottery, jewellery, soaps and sachets. Gloves, scarves and hats. Tweed by the yard, wool on the cone, cards and books
Prices: Under £5 to over £350
Notes: Highland Design and Business Award winner. Finalist in Finest Knitwear Store in UK 1996. Also branch at Armadale on Isle of Skye

SEAN KINGSLEY CERAMICS
The Log Cabin
14 Spring Valley Gardens
Edinburgh
EH3 7NG
Tel: 0131 447 5795
Contact: Sean Kingsley
Established: 1991
Open: 10am to 5.30pm Monday to Friday. 10am to 5pm Saturday. 11am to 5pm Sunday
Craftworkers on site: Yes
Crafts: Domestic earthenware pottery including ovenware and kitchenware. Sculptural ceramics
Prices: £3 to £130. Credit cards not accepted
Note: Disabled access

SIMPLY SCOTLAND
299 Canongate
The Royal Mile
Edinburgh
EH8 8BD
Tel: 0131 557 4056
Contact: Caroline Alderson
Established: 1996
Open: 10am to 5.30pm seven days
Stock: Fine quality hand finished craftwork and designer knitwear. Ceramics, leather, wood, jewellery, textiles, glass, silk, etc. Over 90% of products Scottish made
Prices: £1 to £1100. Visa/MasterCard accepted

THE ADAM POTTERY
76 Henderson Row
Edinburgh
EH3 5BJ
Tel: 0131 557 3978
Contact: Janet Adam
Established: 1983
Open: 10am to 6pm Monday to Saturday
Craftworkers on site: 5
Crafts: Ceramics and handthrown stoneware by Janet Adam. Wide range of ceramics by Emma Hollands, Lucinda McFerran, Michele Bills and Emma Grove
Prices: £5 to £500. No credit cards
Note: Festival Fringe exhibition held. Open workshops where work can be seen being made. No disabled access

THE MEADOWS POTTERY
11A Summerhall Place
(Top of Causewayside)
Edinburgh
EH9 1QE
Tel: 0131 6624064
Established: 1988
Open: 10.30am to 5.30pm Monday to Saturday. Closed 25 December to 7 January
Craftworkers on site: Yes
Crafts: High-fired oxidised stoneware. A wide range of decorative and domestic ware
Prices: From £3 to £160. Credit cards not accepted

THE MULBERRY BUSH
77 Morningside Road
Edinburgh
EH10 4AY
Tel: 0131 447 5145
Contact: Jean Duncan
Established: 1976
Open: 9.30am to 12.30pm and 1.30pm to 5pm daily. Wednesday 9.30am to 12.30pm and 1.30pm to 3.45pm
Stock: Pottery, willow baskets, stained glass, wooden stools, lamp bases and turned bowls, candles, cards, craftmaking supplies
Speciality: High quality wooden toys, e.g. prams, rocking horses, bricks, dolls houses
Prices: £1 to £90. Credit cards not accepted
Note: Many toys made in own craft workshops. As part of Garvald Centre, Edinburgh, we provide craft and work training for young adults with learning disabilities

THE OPEN EYE GALLERY
75/79 Cumberland Street
Edinburgh
EH3 6RD
Tel: 0131 557 1020
Contact: Thomas Wilson
Established: 1983
Open: Monday to Friday 10am to 6pm
and Saturday 10am to 4pm
Stock: High quality domestic pottery,
studio ceramics, glass, wood, jewellery
and sculpture
Prices: From £10 to £500. Credit cards
not accepted
Note: Over twenty contemporary craft
exhibitions held each year

THE SCOTTISH GALLERY
16 Dundas Street
Edinburgh
EH3 6HZ
Tel: 0131 558 1200
Fax: 0131 558 3900
Contact: Amanda Game
Established: 1842
Open: Monday to Friday 10am to 6pm
and Saturday 10am to 4pm
Stock: High quality domestic pottery,
studio ceramics, glass, wood, metal,
jewellery textiles and modern Scottish
paintings
Prices: £12 upwards. Credit cards
accepted
Note: Monthly exhibitions plus wide stock

OATHLAW POTTERY
Oathlaw
Forfar
Angus, DD8 3PQ
Tel: 01307 850272
Contact: Ian or Maggie Kinnear
Established: 1987
Open: 9am to 5pm April to September.
Sunday 2pm to 5pm. Closed Tuesdays.
October to March by appointment
Crafts: Handthrown and individual pieces
in stoneware and raku
Prices: £2 to £200. Credit cards not
accepted
Note: Disabled access

ROGER BILLCLIFFE FINE ART LTD
134 Blythswood Street
Glasgow
G2 4EL
Tel: 0141 332 4027
Fax: 0141 332 6573

Contact: Lynn Park/Roger Billcliffe
Open: Monday to Friday 9.30am to
5.30pm and Saturday 10am to 1pm
Stock: High quality domestic pottery,
glass, wood, metal, jewellery with
modern paintings and sculpture
Prices: £25 to £5,000. Credit cards not
accepted
Note: Monthly exhibitions of applied arts
held here, one of Glasgow's leading
galleries of contemporary paintings and
sculpture. Crafts Council selected gallery

RONALD C.F. BAIRD
9 Old Church Gardens
Bargeddie
Glasgow
G69 7JS
Tel: 0141 781 1129
Contact: Ronald C.F. Baird
Established: 1994
Open: By appointment
Craftworkers on site: 1
Crafts: Assembly of gemstone jewellery
Prices: £1.50 to £5.50. Credit cards not
accepted
Note: No disabled access

THE LANSDOWNE COLLECTION
399 Great Western Road
Kelvinbridge
Glasgow
G4 9HY
Contact: Elisabeth Viguie-Culshaw
Established: 1992
Open: Wednesday to Saturday, noon to
5pm or by appointment
Craftworkers on site: 5
Crafts: Papier mâché frames, mirrors,
hand decorated tinware and terracotta
and stencils. Sponged china, dried
flowers, clocks
Prices: 37p to £200. Visa/Access/Switch
Note: Crafts courses in stencilling, papier
mâché, découpage. Commissions
undertaken

THE SCOTTISH CRAFT CENTRE
The Courtyard
Princes Square
Glasgow
G1 7JN
Tel: 0141 248 2885
Contact: David MacAskill
Established: 1992
Open: Monday to Saturday 10am to 7pm.
Sunday 12 noon to 5pm

Stock: Huge range of quality crafts all made in Scotland by expert craftsmen and women
Prices: £5 to £3,000. Visa/Mastercard/Amex/Switch
Note: Constantly changing range which introduces new craftsmen on a regular basis

HOPSCOTCH
27 West Princes Street
Helensburgh
Dunbartonshire G84 8TF
Tel: 01436 676653
Contact: Ann or Phil Hopkinson
Established: 1989 at previous premises (Caple Pottery). Hopscotch established 1996
Open: 9am to 5pm Monday to Saturday (not seasonal)
Stock: Ceramics, sculpture, prints, textiles, original paintings by local artists, silver jewellery, Mull silver, cushions, candles, cards, dolls houses and accessories
Prices: 95p to £500. All major credit cards accepted
Note: Located in Helensburgh town centre about 50 yards from railway station. Easy access for disabled

CENTAUR DESIGN & HANDCRAFTS
The Groom's Bothy
Traquair House Estate
Traquair
Innerleithen
Peeblesshire
EH44 6PN
Tel: 01896 831376
Contact: Terence Plummer and Deborah Green
Established: 1988
Open: April, June, September, October 12.30pm to 5.30pm. July and August 10.30am to 5.30pm
Craftworkers on site: 2
Crafts: Leatherwork, belts, accessories, including pressed designs. Celtic themed etched jewellery, brass, stainless steel, sterling silver. Kilt belts, celtic buckles, gemstones
Prices: £1 to £100. Visa/MasterCard/Access/Delta/Eurocard
Note: Wholesale orders welcome. Tearoom, grounds, maze. Three other craftworkers on site and all may be seen at work

"TREE TOPS" WOOD WORKSHOP
Harestanes Courtyard, Ancrum
Jedburgh
Roxburghshire TD8 6UQ
Tel: 01835 830742/01450 870212
Contact: Geoff and Lesley Barnard
Established: 1996
Open: 10am to 5pm daily (not Sunday). 7 days from 1st April to end October
Crafts: Wooden toys, mobile shapes, templates, small furnishings, rocking horses, rocking chairs all designed and made in workshop
Speciality: Traditional "cracket" stools
Prices: £1 to £200. Credit cards not accepted
Note: Reclaimed and recycled timber used. Situated in woodland with visitor centre, teashop, children's park & games room, organised walks, other craft studios. Disabled access to visitor centre

JOHN O'GROATS POTTERY & GALLERY
3 Craft Village (harbour car park)
John O'Groats
Caithness KW1 4YR
Tel: 01955 611284
Contact: David & Sally Body
Established: 1977
Open: 9.30am to 5.30pm. 9.30am to 4.30pm Monday to Saturday in winter
Craftworkers on site: 3
Crafts/stock: Handthrown tableware, handbuilt individual ceramics and slip-cast, modelled gift items. Print and card gallery featuring prints of oil paintings by David Body (some originals) and other select artists
Prices: £1 to £300. Visa/MasterCard/Switch/Delta accepted
Note: Disabled access

KINSMAN-BLAKE CERAMICS
Barn House, Smailholm
Kelso
Roxburghshire TD5 7PX
Tel: 01573 460666
Contact: Linda & Rankin Kinsman-Blake
Established: 1978
Open: 10am to 5pm 7 days a week
Craftworkers on site: 2
Crafts: Ceramics – hand thrown and painted majolica, domestic ware. Handmade tiles
Prices: £2 to £200+
Note: Work can be seen in progress. Wheelchair access

KINNAIRD POTTERY
Bankfoot Cottage
Kinnaird by Inchture
Perthshire PH14 9QY
Tel: 01828 686371
Contact: Jane Woodford
Established: 1985
Open: Weekends and summer evenings
till 9pm
Craftworkers on site: 1
Crafts: Decorative hand thrown stone-
ware pottery for the kitchen and dining
table. Also, lamps, vases and lustreware
Prices: £5 to £50. Credit cards not
accepted
Note: Commissions undertaken

JO GALLANT
Ironstones
70 High Street
Kirkcudbright
Kirkcudbrightshire DG6 4JL
Tel: 01557 331130
Contact: Jo Gallant
Established: 1993
Open: 10am to 5pm Monday to Saturday
throughout the summer (except first 2
weeks in July). Other times by appoint-
ment
Craftworkers on site: 1
Crafts: Wall-hangings and cushions
Prices: £20 to £800
Note: Commissions undertaken.
Wheelchair access

OLD MILL POTTERY
Millburn Street
Kirkcudbright
Kirkcudbrightshire DG6 4EB
Tel: 01557 330468
Contact: Wilson Lochhead
Open: 9.30am to 1pm and 2pm to 5pm
Monday to Saturday
Craftworkers on site: 1
Crafts: Domestic pots in both earthen-
ware and stoneware
Prices: £2 to £60
Note: Commissions undertaken

MEARNSCRAFT
5 Cumberland Close
Kirriemuir
Angus DD8 4EF
Tel: 01575 575950
Contact: David Thorpe
Established: 1996
Open: Tuesday to Saturday 10am to 5pm

Craftworkers on site: 2
Crafts: Embroidery kits of maps of
Scottish islands and counties
Prices: £2 to £30+. Visa/Access/
American Express etc.
Note: Visitors are welcome to watch
work in progress

KNOCKAN STUDIO
Knockan
Elphin by Lairg
Lairg
Sutherland
IV27 4HH
Tel: 01854 666261
Fax: 01854 666261
Contact: Linda Combe
Established: 1994
Open: 9am to 6pm April to October.
Closed Sundays. Other times by appoint-
ment
Stock: Jewellery, leaded glass, knitwear,
pottery, ceramics, woodwork, textiles and
general crafts. All produced in Scotland to
a high standard
Speciality: Jewellery incorporating
Scottish gemstones. Leaded glass
Prices: £1 to £2,000. Visa/Access
Note: Workshop area visible to public

HIGHLAND STONEWARE
Baddidaroch
Lochinver
Sutherland
IV27 4LP
Tel: 01571 844376
Contact: David Grant
Established: 1974
Open: 9am to 6pm weekdays (Saturdays
seasonal)
Craftworkers on site: 20
Crafts: Huge ranges of freehand-painted
tableware and giftware in high-fired
stoneware
Prices: From under £5 to over £200

REDFERN ORIGINALS
Coul of Fairburn
Marybank
by Muir of Ord
Ross-shire
IV6 7QD
Tel: 01997 433369
Contact: Chris or Wanda Redfern
Established: 1991
Open: 10am till late. Closed Tuesdays
and Thursdays

Craftworkers on site: 2
Crafts: Hand screen-printed greetings cards and stationery. Photographic cards
Note: First floor studio – no wheelchair access. Bed and breakfast available

THE STATION GALLERY
Melrose Station
Palma Place
Melrose
Roxburghshire
TD6 9PR
Contact: Bruce Dobson
Established: 1995
Open: 10am to 5pm daily (closed Mondays in winter)
Craftworkers on site: 1
Crafts: Wood sculptures, ceramics, metal sculpture, pottery, prints, cards etc. Monthly guest artist exhibitions
Prices: £4 to £900. Access/Visa/Delta/ Mastercard/Switch/Eurocard
Note: Picture framing, art commissions, working artist in house

THE McPHAIL COLLECTION
Camp Cottage
Selkirk Road
Moffat
Dumfriesshire
DG10 9LF
Tel: 01683 220074
Contact: John McPhail
Established: 1985
Open: Most days please 'phone first
Craftworkers on site: 2
Crafts: Sculptured metal figures, clocks, vases, bookends, candleholders
Prices: £7 to £200

CLAIRE WHITE STUDIOS
Unit 15, Enterprise House
Dalziel Street
Motherwell
Lanarkshire ML1 1PJ
Tel: 01698 327070
Contact: Claire White
Established: 1993
Open: Monday to Friday 10.30am to 4.30pm
Craftworkers on site: 2
Crafts: Needlecraft kits, watercolour panels and cards. Scottish designer craftwork
Prices: £2.95 to £175. Visa/Mastercard
Note: Visitors travelling a long distance are advised to telephone first

HIGHLAND POTTERY
Church Terrace
Newtonmore
Inverness-shire
PH20 1DT
Tel: 01540 673636
Contact: William Forrest
Open: 10am to 5pm daily Easter to October
Crafts: Handmade pottery, stoneware, domestic and giftware items, individually hand-decorated

EDINBURGH CRYSTAL VISITOR CENTRE
Eastfield
Penicuik
Midlothian EH26 8HB
Tel: 01968 675128
Contact: Sandra Campbell
Open: Monday to Saturday 9am to 5pm. Sunday 11am to 5pm
Crafts: Handblown, fully cut, engraved lead crystal. Extensive range includes trinket boxes, tumblers, clocks, candlesticks, decanters, bowls, vases and wine suites. Amex/Visa/MasterCard accepted
Note: Factory tours (adult £2 child £1) Monday to Friday 9am to 3.30pm. Saturday and Sunday May to September 11am to 3pm. Located 10 miles south of Edinburgh on the A701 to Peebles.

CAITHNESS GLASS VISITOR CENTRE
Inveralamond Industrial Estate
Perth
PH1 3TZ
Tel: 01556 600200
Open: Monday to Saturday 9am to 5pm. Sunday 11am to 5pm. Sunday October to March 9am to 5pm
Crafts: Art glass paperweights, tableware. Factory shop with viewing gallery of glass making area. Collectors gallery/museum
Note: Glassmaking Monday to Friday 9am to 4.30 pm. Located just off roundabout on the A9 north side of Perth

POTS OF PITLOCHRY
Mill Lane
Pitlochry
Perthshire PH16 5BH
Tel: 01796 474367
Open: 10am to 5pm summer, 11am to 4pm winter
Crafts: Handthrown domestic stoneware, decorated earthenware, figurative sculpture and visual arts

THE GALLERY
East Haugh
Pitlochry
Perthshire
PH16 J5
Tel: 01796 472147
Contact: Kate Johnston
Open: 10am to 6pm
Stock: Original pictures and gifts. Unique collection of hand painted mirrors and frames. Footstools, paintings, pottery, jewellery, etc. All made in Scotland

THE GINGERBREAD HORSE
9 High Street
Pittenweem
Fife
KY10 2LA
Tel: 01333 311495
Contact: Hazel Moodie
Established: 1977
Open: 9.30am to 5.30pm Tuesday to Saturday. 12 noon to 5.30pm Sunday
Crafts/stock: Prints, jewellery, baskets, ceramics
Prices: £1 to £250. Visa/Access
Note: Tea/coffee shop with home baking. Patio open in summer for coffees. St Fillan's Cave

ALISON THOMSON
30 High Street
Sanquhar
Dumfriesshire
DG4 6BL
Tel: 01659 58264
Contact: Alison Thomson
Open: Monday to Friday 10am to 5pm (closed 1pm to 2pm). Half day Thursday and Saturday
Stock: Original Sanquhar Jerseys, stockings, hats, mitts and jackets. Sanquhar has a fine and famed tradition in knitwear and patterns that date back to the 16th century
Prices: £12 to £69

SELKIRK GLASS
Dunsdle
Hough
Selkirk
TD7 5FS
Tel: 01750 20954
Contact: Ron Hutchinson
Established: 1977
Open: Monday to Friday 9am to 5pm. Saturday 10am to 4.30pm. Sunday 12 noon to 4pm
Craftworkers on site: 8
Crafts: Handmade art glass paperweights and vases
Prices: From £7 upwards. Major credit cards accepted
Note: Glass making Monday to Friday 9am to 4.30pm. Located on the A7 on the north side of Selkirk

PETER GASPAR DESIGN
Glenkiln
Shawhead
Dumfriesshire
DG2 9UE
Tel: 01387 730397
Contact: Peter Caspar
Established: 1993
Open: Please telephone for details
Stock: Designer knitwear, garden chairs, deco lighting and furniture
Prices: £15 to £500. Access/Visa/Mastercard

CHURCH SQUARE CERAMICS
4 Church Square
St. Andrews
Fife
KY16 9NN
Tel and Fax: 01334 477744
Contact: George or Julia Young
Established: 1984
Open: 9.30am to 5.30pm Monday to Saturday. Sunday seasonal
Craftworkers on site: 1
Crafts/Stock: Wide range of hand thrown decorative, domestic stoneware and porcelain. Ceramic earrings
Prices: From under £5 to over £50

SIMPLY SCOTLAND
158 South Street
St. Andrews
Fife
KY16 9EQ
Tel: 01334 477651
Contact: Caroline Alderson
Established: 1990
Open: 10am to 5.30pm Monday to Saturday and Sunday July, August. Closed Thursdays January, February, March
Stock: Fine quality hand finished craftwork and designer knitwear. Ceramics, leather, wood, jewellery, textiles, glass, silk, etc. Over 90% of products Scottish made
Prices: £1 to £1100. Visa/MasterCard

THE POTTERY WORKSHOP
57A South Street
St. Andrews
Fife KY16 9QR
Tel: 01334 472985
Contact: Ian Wood
Established: 1987
Open: 10am to 5pm weekdays. 12 noon to 4pm Sundays. Closed Thursdays in winter
Craftworkers on site: 3
Crafts: Decorated stoneware, coloured porcelain, tiles and experimental one-off pieces by Anne Lightwood and Ian Wood
Prices: £2 up to £400 Most major credit cards

CELTIC LEGEND
The Swinton Pottery, 25-27 Main Street
Swinton
Duns, Berwickshire TD11 3JJ
Tel: 01890 860283
Contact: John or Sandra Day
Established: 1984
Open: 10am to 5pm Tuesday to Saturday
Craftworkers on site: 12
Crafts: Hand modelled fantasy figures. Hand thrown pottery
Prices: £1.50 to £100. Credit cards not accepted
Note: Viewing window into workshop. Wheelchair access

THE TAIN POTTERY
Aldie
Tain
Ross-shire IV19 1LZ
Contact: Robert Hudson or Alen Pett
Established: 1996
Open: 10am to 5pm (trade 8am to 5.30pm)
Craftworkers on site: 5
Crafts: Local weaving, woodturning and basket weaving
Prices: £5 to £100. Access/Visa/MasterCard
Note: Working watermill, open workshop, tourist information

ANDREW WEATHERHEAD POTTERY
The Old Smithy
Kirkland, Moniaive
Thornhill
Dumfriesshire DG3 4HD
Tel: 01848 200548
Established: 1994
Open: 10am to 5pm daily
Stock: A wide range of decorative, functional and lettered earthenware pottery and tiles.
Speciality: Brush decoration with both Islamic and Celtic influences
Prices: £5.00 upwards. All major credit cards accepted

WOOD 'N THINGS
Gravel Pit Cottage
Kirkland, Moniaive
Thornhill
Dumfriesshire DG3 4HB
Tel: 01848 200345
Contact: F & D Berry
Established: 1984
Open: 10am to 5pm summer. 10am to 4pm winter. Closed Tuesday morning and all day Thursday
Crafts: Small wooden wheeled toys, cars, lorries, aircraft, turned wood articles, bowls, boxes, candlesticks etc.
Prices: £1 to £40

GLASS CREATIONS
Thurso Glass Studios, Riverside Road
Thurso
Caithness
Tel: 01847 894017/895637
Contact: Maureen Pearson
Established: 1990
Open: 10am to 5pm. Closed Mondays and Sunday morning
Crafts: Lampworked glass for all occasions ranging from animals to candlesticks to sculpture
Prices: £5 to £5000. Access/Visa/MasterCard accepted
Note: Demonstrations of glass making Monday/Wednesday/Thursday 7pm to 10pm, Saturdays 10am to 5pm. Access for wheelchairs

TWEEDSMUIR GLASS
The Crook Inn
Tweedsmuir
near Biggar, Lanarkshire ML12 6QN
Tel: 01899 880282
Fax: 01899 830466
Contact: Chris Dodds
Established: 1992
Open: 9am to 5pm
Craftworkers on site: 1
Crafts/stock: Glass animals, vases, paperweights etc. Wood, china, cards
Prices: £5 to £50. Credit cards not accepted
Note: Demonstrations held

URQUHART POTTERY
School House
Urquhart
by Elgin
Morayshire IV30 3LG
Tel: 01343 842464
Contact: Clare Smith
Open: 9am to 5pm
Craftworkers on site: 1
Crafts: Oxidised stoneware, thrown on the wheel and decorated with sprig moulds of fish, flowers and animals
Prices: £3 to £70

CAITHNESS GLASS VISITOR CENTRE
Airport Industrial Estate
Wick
Caithnesss KW1 5BP
Tel: 01955 602286
Contact: Mrs. K. Florence
Open: Monday to Saturday 9am to 5pm. Sunday (Easter to October 31) 11am to 5pm
Crafts: Handmade glass artware, engraved glass and jewellery
Note: Glassmaking demonstrations Monday to Friday 9am to 4.45pm

NORTH SHORE CERAMICS
4B Airport Industrial Estate
Wick
Caithness KW1 4QS
Contact: Jenny Mackenzie-Ross
Established: 1993
Open: 9.30am to 5pm Monday to Friday. Saturdays during July and August
Craftworkers on site: 1
Crafts: Hand thrown and decorated stoneware pottery
Prices: £4 to £120. Visa/MasterCard
Note: Work can be seen in progress

THE ISLANDS

SILK & STAINED GLASS STUDIO
Kinloch Hotel, Blackwaterfoot
Isle of Arran
KA27 8EP
Tel: 01770 860444
Fax: 01770 860417
Contact: Alison Bell
Established: 1996
Open: 10am to 12.30pm and 1.30pm to 5pm
Craftworkers on site: 1-2

Crafts/stock: Hand painted silk accessories, wallhangings, cards, ties, scarves, waistcoats. Stained glass, lamps and mirrors etc.
Prices: £5 to £900. Visa/Access/MasterCard
Note: Commissions undertaken. Silk painting and stained glass classes held daily by appointment.

CHRISTINA MACLEOD
Rockview
Scadabay
Isle of Harris
HS3 3ED
Tel: 01859 511215
Contact: Christina Macleod
Open: 9am to 8pm Monday to Saturday all year
Craftworkers on site: 1
Crafts: Harris knitting yarn, Harris Tweed and knitwear

CLO MOR
1 Liceasto
Isle of Harris
HS3 3EL
Tel: 01859 530364
Contact: Anne Campbell
Open: 9am to 5pm Monday to Friday (May to September inc.)
Craftworkers on site: 1
Crafts: Hand spun weft Harris Tweed; naturally dyed, handspun, handwaulked tweeds woven on traditional wooden loom

CROFT CRAFTS HARRIS
4 Plockropool
Drinnishader
Isle of Harris
HS3 3EB
Tel: 01859 511217
Contact: Katie Campbell
Open: 9am to 7pm Monday to Saturday all year
Craftworkers on site: 1
Crafts: Harris Tweed, knitting wools and knitwear
Note: Weaving demonstrations held

FLORA ANN MACLEOD
Harmony Villa
Scadabay
Isle of Harris
HS3 3ED
Tel: 01859 511221

Contact: Flora Ann Macleod
Open: 9am to 8pm Monday to Saturday
April to October
Craftworkers on site: 1
Crafts: Hand woven Harris Tweed, hand
knitted goods, knitting wool, etc.

LUSKENTYRE HARRIS TWEED CO
6 Luskentyre
Isle of Harris
HS3 3HL
Tel: 01859 550261
Contact: D.J. and M.D. Mackay
Established: 1992
Open: 9am to 6pm Monday to Saturday
all year
Craftworkers on site: 1
Crafts/Stock: Handwoven Harris Tweed;
specialising in Tartan; also hats, ties,
scarves, jumpers
Prices: From 17.50 to £45.00. Major
credit cards accepted

ROSE COTTAGE INDUSTRIES
Rose Cottage
Tarbert
Isle of Harris
Tel: 01859 502226
Contact: Francine Johnson
Established: 1995
Open: 10am to 5pm all year
Craftworkers on site: 2
Crafts: Quality handmade Lewis chess
sets and tile boards; hand painted
selection of collectors' chess sets
Prices: £50 to £300. Major credit cards
accepted

SOAY STUDIO
West Tarbert
Isle of Harris
Tel: 01859 502361
Contact: Margaret Mackay
Open: 9am to 12.30pm Monday to Friday
May to September
Craftworkers on site: 1
Crafts: Traditional natural dyeing,
handspinning; knitted, woven and
embroidered items for sale, also natural
dyed wools

IONA POTTERY
Burnside Cottage
Isle of Iona
Argyll
PA76 6SW
Tel: 01681 700439/700328

Contact: Gordon Menzies
Established: 1982
Open: 9am to 5pm April to September.
By appointment October to March
Craftworkers on site: 1 (sometimes 2)
Crafts: Decorative thrown and handbuilt
stoneware
Prices: £5 to £400. Major credit cards
accepted
Note: Gallery with paintings of Iona and
original prints

IONA SCOTTISH CRAFTS
Isle of Iona
Argyll
Tel and Fax: 01852 300434
Contact: Manageress 01681 700332
Established: 1964
Open: 9.30am to 5.30pm March to
beginning October, Monday to Saturday.
Open 7 days June, July and August
Stock: Pottery, knitwear, jewellery, silver
and pewter, cards etc.
Prices: 90p to £160. Visa/MasterCard

ELIZABETH SYKES BATIKS
Port Charlotte
Isle of Islay
Argyll
PA48 7UD
Tel: 01496 850357
Fax: 01496 850357
Contact: Liz Sykes
Established: 1982
Open: 11am to 5pm 7 days a week
Craftworkers on site: 2
Crafts: Hand-dyed batiks produced by
owner. Greetings cards and prints
Speciality: Original silk batik pictures,
scarves and garments. Prints & frames etc.
Prices: £1 to £400. Visa/Access/Delta/
Mastercard/Eurocard
Note: One day introductory batik
workshops held throughout the year. 70
year-old letterpress printing press. Visitors
are advised to wear wellies when wet

BACK STREET GALLERY
15 Swainbost
Ness
Isle of Lewis
HS2 0TA
Tel: 01851 810295
Contact: Donna Scott
Open: 10am to 5pm all year
Stock: Varied range of arts and crafts from
North West Lewis

BOGHA FROIS KNITWEAR
5 Shulishader, Point
Isle of Lewis
Tel: 01851 870715
Contact: Jean Staines
Open: 2pm to 6pm Monday to Saturday
all year
Craftworkers on site: I
Crafts: Machine knitwear in various
yarns, original design sweaters including
maps, Lewis chessmen pattern
Note: Spinning demonstrations held

BOIREIG KNITWEAR
28 Outend
Swordale, Point
Isle of Lewis
Tel: 01851 870128
Contact: Erica Morrison
Open: 10am to 6pm all year
Craftworkers on site: 1
Crafts: Knitwear specialising in chunky
knits for the 15-30 age market

BOREAS KNITWEAR
29A Breasclete
Isle of Lewis
HS2 9EF
Tel: 01851 621241
Contact: Barry and Sheila Leigh
Open: 10am to 8pm all year
Craftworkers on site: 2
Crafts: Machine and hand knits in Harris
and other yarns

BORGH POTTERY
Fivepenny House
Borgh
Isle of Lewis
HS2 0RX
Tel: 01851 850345
Contact: Alex and Sue Blair
Established: 1978
Open: 9.30am to 6.30pm all year
Craftworkers on site: 2
Crafts: Handthrown domestic stoneware
and porcelain. Gas fired with a variety of
glazes. Some handmade terracotta garden
pots.
Prices: £5 upwards
Note: Commissions accepted

CO-CHOMUNN NA PAIRC
Tigh Sgoile
Kershader
Lochs
Isle of Lewis

Tel: 01851 880236
Contact: Ina Mackinnon
Open: 11am to 5pm Monday to Saturday
all year
Craftworkers on site: 1
Crafts: Hand knitted Harris yarn sweaters
and hose
Note: Cafeteria, shop, hostel accommo-
dation and Field Study Centre

FEAR AN EICH
Coll Pottery, Back
Isle of Lewis
HS2 0JP
Tel: 01851 820219/565
Contact: G.B. and C.E. Edwards
Open: 9am to 6pm Monday to Saturday
all year
Craftworkers on site: 2
Crafts: Wide range of ceramics (including
island figures, animals and birds) & crafts

GIFTS UNLIMITED
9 Bayhead Street
Stornoway
Isle of Lewis
Tel: 01851 703337
Contact: Beatrice M. Schulz
Open: 9.30am to 5.30pm Monday to
Saturday all year
Stock: Wide range of local and Scottish
crafts, knitwear, Harris Tweed, the
'Macbear' Family

HARBOUR VIEW
Port of Ness
Isle of Lewis
HS2 0XA
Tel: 01851 810735
Contact: Kate and Anthony Barber
Open: 11am to 5.30pm Monday to
Saturday March to November
Stock: Exclusive island watercolours and
prints, knitwear, books and gifts
Note: Cafe here also

HARRIS KNITWEAR
16 Jameson Drive
Stornoway
Isle of Lewis
Tel: 01851 703265
Contact: Morag Macleod
Open: 9am to 10pm all year except
Sundays
Crafts: Harris wool, hand knitted and
hand framed knitwear. Aran ribs, hose,
hats, etc.

HEATHER GRAHAM CRAFTS
Borve Cottage, Borve
Isle of Lewis
HS3 3HT
Tel: 01859 550202
Fax: 01859 550283
Contact: Heather Graham Scherr
Open: 9am to 6pm Monday to Saturday
all year
Stock: Shetland wool cardigans, jerseys
and waistcoats; also Harris wool products
and island crafts

KELLS
7 James Street, Stornoway
Isle of Lewis
Tel: 01851 705351
Contact: Heather Butterworth
Open: 9am to 5pm Monday to Thursday.
9am to 3pm Friday. All year
Crafts: Harris Tweed waistcoats, scarves,
gifts, etc.

LEWIS LOOM CENTRE
The Grainstore
3 Bayhead, Stornoway
Isle of Lewis
Tel: 01851 703117
Contact: James R. Mackenzie
Open: Monday to Saturday April to
October 10am to 6pm (also some
evenings)
Craftworkers on site: Yes
Crafts: Harris tweed, quality knitwear and
knitting yarns
Note: Guided tours (30-40 minutes) with
demonstrations of the story of Harris
Tweed. Shop

MORVEN GALLERY
Morven, Barvas
Isle of Lewis
HS2 0QX
Tel: 01851 840216
Contact: Janis Scott
Open: 9am to 6pm Monday to Saturday
until mid-October
Stock: Paintings, textiles, ceramics, prints,
photography from the Western Isles
Note: Gallery and cafe

ORINSAY KNITWEAR
7 Orinsay
Lochs
Isle of Lewis
Tel: 01851 880394
Contact: Mrs. Nellie A. Kennedy

Open: 9am to 8pm Monday to Saturday
all year
Craftworkers on site: 1
Crafts: Knitwear in 100% Harris yarn,
Harris Tweed
Note: Weaving demonstrations held

RACHEL MURRAY KNITWEAR
47 Habost , Ness
Isle of Lewis
Tel: 01851 810396
Contact: Rachel Murray
Open: 9.30am to 7.30pm all year
Craftworkers on site: 1
Crafts: Harris wool knitwear

SANDRA M. GILLIES KNITWEAR
Lochview, South Shawbost
Isle of Lewis
Tel: 01851 710419
Contact: Mrs. Sandra Gillies
Open: 9.30am to 8pm Monday to
Saturday all year
Stock: Large selection of Harris wool
knitwear, Harris tweed.
Note: Knitting and weaving demonstra-
tions held

WESTEND COTTAGE CRAFTS
Westend Cottage
Shulishader, Point
Isle of Lewis
Tel: 01851 870616
Contact: John Macmillan
Open: 9am to 6pm Monday to Saturday
March to September
Craftworkers on site: 1
Crafts: Handwoven Harris Tweed,
woollen knitwear, crafts, Harris yarn
Note: Free demonstrations held on
Hattersley and Rapier Loom

ELLI PEARSON
Hestily
Windwick, South Ronaldsay
Orkney
Tel: 01856 831355
Contact: Elli Pearson
Established: 1978
Open: 10am to 6pm Monday to Saturday
May to September. Other times by
arrangement
Craftworkers on site: 2
Crafts: Pottery inspired by Orkney
landscape incorporating rich earthy
browns and vibrant turquoises
Prices: £15 to £75 Visa/MasteCard

FURSBRECK POTTERY
Harray School, Harray
Orkney
KW17 2JR
Tel: 01856 771419
Fax: 01856 771419
Contact: Andrew Appleby
Established: 1976
Open: 9am to 6pm all year
Crafts: Pottery, goblets, jugs, vases, mugs, plates. Orkney crafts. Candles, hats, gloves, jewellery
Prices: 75p to £300. Access/Visa/Mastercard
Note: Craft demonstrations held

HOXA TAPESTRY GALLERY
Neviholm
Hoxa, South Ronaldsay
Orkney
Tel: 01856 831395
Contact: Leila Thomson
Established: 1996
Open: 7 days. 10am to 6pm April to September. 10am to 3.30pm October to March
Stock: Hand-woven tapestries inspired by landscape of Orkney. Samplers and cards
Note: Visitors may see artist at work. Weaving courses held. Wheelchair access

HROSSEY SILVER
Crumbecks
Petertown Road
Orphir
Orkney
Tel: 01856 811347
Contact: Peter and Jaen Rowland
Open: 10am to 3pm Tuesday to Saturday. Other times by arrangement
Crafts: Designs combining handmade and cast jewellery with larger items, e.g. scent bottles. Every item hallmarked in Edinburgh and comes complete with own presentation box
Prices: £25 upwards. Visa/MasterCard/Access

ORKNEYINGA
Holland Cottage
Marwick
Birsay
Orkney
Tel: 01856 721359
Contact: Elizabeth and Kevin Allen
Established: 1989

Open: 10am to 5pm Monday and Wednesday to Saturday May to October. Other times by arrangement
Craftworkers on site: 2
Crafts: Silver jewellery, bookmarks, candle snuffers and drinking cups
Prices: £18 to £430. Visa/MasterCard
Note: Workshop located on a hill overlooking the sea and two RSPB bird reserves

ORTAK
Hatston Industrial Estate
Kirkwall
Orkney
Tel: 01856 872224
Contact: Liz Myles
Established: 1967
Open: 9am to 5pm Monday to Saturday all year. Sunday is seasonal
Craftworkers on site: Yes
Crafts: Gold and silver jewellery influenced by Norse and Scottish cultures. Variety of stones and metals used. Handcrafted gifts
Prices: £2 to £200. Switch/Visa/Access/MasterCard/Amex
Note: Visitor centre and shop with free parking at Hatston. Video presentation. Demonstrations of jewellery being hand-crafted

SCAPA CRAFTS
12 Scapa Court
Kirkwall
Orkney
Tel: 01856 872517
Contact: Jackie and Marlene Miller
Established: 1993
Open: 10am to 5pm Monday to Saturday April to October
Crafts: Traditional Orkney chairs
Prices: £60 to £750
Note: Visitors are able to see traditional straw backs being made.

STEPHANE JAEGER
Littlequoy, Burray
Orkney
Tel: 01856 731228
Contact: S. Jaeger
Established: 1989
Open: 10am to 6pm Monday to Saturday June to September (Sundays by arrangement)
Craftworkers on site: 1
Crafts: Hand-spun and hand-knitted

sweaters and other knitwear
Prices: £15 to £250 Visa/Mastercard/ Eurocheque
Note: Hand-spinning demonstrations

*TAIT & STYLE
Brae Studio
Old Academy, Stromness
Orkney
Tel: 01856 851186
Contact: Ingrid Tait
Established: 1980
Open: 11am to 1pm and 2pm to 4pm Monday to Friday. 10am to 1pm Saturday May to September
Stock: Scarves, throws, cushion covers etc featuring techniques such as fabric-fusion and dip-dyeing combined with traditional embroidery and felting on tweeds, wools and tartans; waistcoats and hats
Prices: £20 to £250. Visa/MasterCard

TANGLE DESIGNS
Tangles, Stenness
Orkney
Tel: 01856 851065
Contact: Mary Smith
Open: 10.30am to 5pm Monday to Saturday Easter to September
Crafts: Soft fabric hats from velvet and tweed. Felt waistcoats and bags
Prices: £2.50 to £150 Visa/MasterCard

THE QUERNSTONE
38 Victoria Street, Stromness
Orkney
KW16 3AA
Contact: Elaine Bentley
Established: 1984
Open: Summer 9am to 5pm, except Sundays 10.30am to 3.30pm, plus evenings 7.30pm to 9.30pm Friday and Saturday
Stock: Hand framed knitwear and hand knitted garments. Jewellery, ceramics, gifts
Prices: £1 to £100

THE WORKSHOP
Back Road
St Margaret's Hope, South Ronaldsay
Orkney
Tel: 01856 831587
Established: 1978
Open: Monday to Saturday: 10am to 5pm April to December. 10am to 1pm January to March. Sundays by arrangement
Craftworkers on site: Yes

Craft: High quality knitwear from original designs inspired by traditional fishermen's patterns including Arran, Icelandic Lopi and wool/silk mix
Prices: £79 to £92.50 Visa/MasterCard/ Travellers' cheques

TRADITIONAL ORKNEY CRAFTS
Mariveg, Rope Walk
Kirkwall
Orkney
Tel: 01856 875110
Contact: Arthur and Elizabeth Eccles
Open: 9am to 5pm Monday to Friday all year. Other times by arrangement
Crafts: Quality furniture including traditional Orkney chairs. Stools, nests, dining stools and footstools. Range of stools and tables based on designs of Charles Rennie Mackintosh
Prices: £35 to £680. Visa/MasterCard/ Eurocard/Access

SHETLAND WORKSHOP GALLERY
4-6 Burns Lane
Lerwick, Zetland
Shetland Islands
ZE1 0EJ
Tel: 01595 693343 **Fax:** 01950 477550
Contact: Gill Finnie
Established: 1975
Open: 9.30am to 1pm and 2pm to 5pm
Stock: Wild life figurines, exclusive knitwear plus complementary range of clothes, rugs, cushions and glass
Speciality: Figurines/knitwear collected worldwide
Prices: £1 to £100. Most major cards
Note: Located in one of Lerwick's oldest most picturesque lanes. Regarded by many as one of the most interesting gift/ craft shops in the north of Scotland

BORRERAIG PARK
Borreraig Park, By Dunvegan
Isle of Skye
IV55 8ZX
Tel: 01470 511311
Contact: Audrey M. Spencer
Established: 1989
Open: 10am to 6pm daily, all year
Stock: Quality knitwear and other crafts made locally. Kilt and climbing hose, walking socks, fleeces, wool, walnut needles, silver and much more
Prices: £1 to £200 MasterCard/Visa/ Eurocard accepted

CARBOST CRAFT POTTERY
Carbost Mhor
Carbost
Isle of Skye
Tel: 01478 640259
Contact: Geoff & Judith Nicolls
Open: Pottery 10am to 7pm, tearoom
11am to 5pm; seven days a week
Crafts/stock: Extensive range of items
handcrafted on the premises including
unique vases, tableware and giftware.
Wide range of sterling silver jewellery

CROFT STUDIO
Dunvegan
Isle of Skye
IV55 8GT
Tel: 01470 521383
Contact: Donald Budge
Established: 1970
Open: 9.30am to 5.30pm Monday to
Saturday. 10.30am to 4.30pm Sunday
Stock: Original paintings, prints. Celtic
designed wall hangings, scarves and table
linen. Lampshades decorated with Skye
flora
Speciality: Handpainting on fabric and
wood
Prices: £1 to £1,500. Visa/Mastercard
Note: 95% of work produced by Daisy,
Pamela and Donald Budge. Designs
inspired by ancient celtic manuscripts
and the beauty of Skye. Tea and coffee
available November to March

DANDELION DESIGNS & IMAGES GALLERY
The Captain's House
Stein
Waternish
Isle of Skye
Tel: 01470 592218/223
Contact: Cathy Myhill
Established: 1993
Open: Daily from 11am
Stock: Hand-decorated woodcrafts,
clocks and boxes. Original paintings,
photographs, handmade cards, jewellery,
clothing, books and postcards. Most
made in Skye and own workshop
Visa/MasterCard/American Express
Note: Easy access for disabled

DUNHALLIN CRAFTS
Dunhallin
Waternish
Isle of Skye
Tel: 01470 592271
Contact: Margaret MacKinnon
Open: Daily during holiday season
Crafts: Handmade top quality knitwear
for all the family

EDINBANE KNITWEAR
Lochview
Edinbane
Isle of Skye
Tel: 01470 582310
Open: Throughout the year 9am to 9pm.
7 days in summer
Crafts/stock: Quality heavy hand knitted
traditional Arrans, Harris wool hand and
machine knits. Harris Tweed from the
roll. Mohair knitwear, sheepskin goods
and other local crafts
Note: See knitting machines working

RAGAMUFFIN
Armadale
Sleat
Isle of Skye
IV45 8RS
Tel: 01471 844 217
Established: 1975
Open: 9am to 6pm or later, seven days a
week
Stock: Designer knitwear from Britain and
Ireland, Fair Isles, chunkys, handknits and
Shetlands. Our own designs in tweeds
and silks. Pottery, jewellery, soaps and
sachets. Gloves, scarves and hats. Tweed
by the yard, wool on the cone, cards and
books
Prices: under £5 to over £350
Note: Located on Armadale Pier, 5
minutes from Clan Donald Centre.
Highland Designer and Business Award
winner. Finalist in Finest Knitwear Store
in UK 1996. There is another branch in
Edinburgh

THE UIG POTTERY
Idrigill
Uig, by Portree
Isle of Skye
IV51 9XU
Tel: 01470 542424
Open: 9am to 6pm Monday to Saturday
(7 days in summer season)
Craftworkers on site: Yes
Crafts: Functional individual pieces,
colourful Isle of Skye inspired decorations
on high-fired, reduced porcelainous
stoneware

ALAN J. RADDON SHOEMAKER/ DESIGNER
Clifton
Aberarth
Dyfed
SA46 0LW
Tel: 01545 570904
Contact: Alan Raddon
Established: 1987
Open: Most days but please telephone for details
Craftworkers on site: 1
Craft: Shoes, sandals. High quality comfortable shoes made to last
Prices: £80 to £190+. Credit cards not accepted
Note: Exhibition of shoes and rope soled shoes

ABERYSTWYTH ARTS CENTRE
Penglais Hill
Aberystwyth
Dyfed
SY23 3XDE
Tel: 01970 622895
Fax: 01970 622 883
Open: Monday to Saturday 10am to 5pm. Closed Sundays. Open evenings on performance nights. Closed early June for university exams
Stock: Domestic pottery, studio ceramics, glass, metal, wood, jewellery, textiles
Note: Situated in university arts centre. Regular crafts exhibitions. Programme available on request – specialises in ceramics

PENNAU CRAFT SHOP
Bow Street
near **Aberystwyth**
Dyfed
SY24 5AA
Tel: 01970 820050
Contact: Martin Miles
Established: 1982
Open: 10am to 5.30pm (closed Mondays) – open 7 days a week June/July/August/ September
Stock: Leather products, celtic jewellery, pottery, slate, knitwear, pictures and wide range of Welsh crafts
Prices: 50p to £50. Credit cards not accepted
Note: Building converted from old farm buildings. Coffee shop on site

COLIN PEARCE FURNITURE MAKERS
Unit 1-2
Llys Llywelyn
Aberffraw
Anglesey
Gwynedd
LL63 5AQ
Tel: 01407 840180
Contact: Colin Pearce
Established: 1990
Open: 8am to 5pm Monday to Friday
Craftworkers on site: 1/2
Crafts: Fine quality contemporary furniture
Prices: £1 to £25,000. Access/Visa/ Mastercard etc.
Note: Short and long term courses offered to suit individual needs. Coastal Heritage Centre and Cafe close

FELIN WEN STUDIOS
Rhosgoch
Anglesey
Gwynedd
LL66 0AS
Tel: 01407 831419
Contact: Beverly Belshaw
Established: 1995
Open: Most days but please telephone first
Craftworkers on site: 1
Crafts: Silk paintings, silk scarves, silk ties, accessories, greeting cards
Prices: 75p to £80. Credit cards not accepted
Note: High quality products and customer focus. Demonstrations and workshops at local art centre

SAM GLASS ROCKING HORSES
Bryn Ifor
Brynsiencyn
Anglesey
Gwynedd
LL61 6UG
Tel: 01248 430603
Contact: Sam Glass
Established: 1988
Open: Most days but please telephone first for details
Craftworkers on site: 1
Craft: Mainly traditional wooden rocking horses
Prices: £575 to £20,000. Credit cards not accepted

THE AGGIE WAGGY WORKSHOP
The Library Hall
Ffordd Garnedd
Bangor
Gwynedd
LL57 1NT
Tel: 01286 870899
Contact: Alison Mercer
Established: 1994
Open: appointment advisable
Craftworkers on site: 1
Crafts: Batik and embroidered textile artifacts from jewellery to hangings
Prices: £5 to £3000. Credit cards not accepted
Note: Demonstrations, talks, lectures held. Commissions undertaken. Disabled access

PAMIKA 3 CUSHION DESIGNS
The Old Stable
Pentrafoelas
Betws-y-Coed
Gwynedd
LL24 0TY
Tel: 01690 770603
Contact: Mr. M.F. Johnson
Established: 1978
Open: 10am to 4pm
Craftworkers on site: 3/4
Crafts: Soft furnishings and all accessories. Pillows, feather inners
Prices: £2.99 to £600. Credit cards not accepted
Note: Water mill, teashop and heritage walk

ORIEL Y DDRAIG
Church Street
Blaenau Ffestiniog
Gwynedd
LL41 3HD
Tel: 01766 831777
Contact: Pat Clarke and Peter Elliott
Established: 1984
Open: 9.30am to 5pm Tuesday to Saturday throughout the year
Stock: Craft materials for candlemaking, batik, silk dyeing and cold dyeing, calligraphy, marbling, lino cutting, clay, stone powder, resins etc. Contemporary artists' prints. Artists' cards
Speciality: Wide range of papers. Picture framing on premises
Note: 5 or 6 art exhibitions held each year showing paintings, sculpture, printmaking, tapestry and weaving

HERE BE DRAGONS
The Old School, Llaneglwys
Builth Wells
Powys LD2 3BQ
Tel: 01982 560337
Contact: Barry and Polly Vernon
Established: 1988
Open: By telephone appointment
Craftworkers on site: 2
Crafts: Dragons and many other fantasy figures and teddies
Prices: 95p to £95. Visa/Mastercard
Note: Working craft shop with ceramics in all stages of production. Cromartie dealer – tuition and supplies. Refreshments free. Wheelchair access – but not to toilets. Member of the Wales Craft Council

CLAI
1 The Turrets
Llanrug
Caernarfon
Gwynedd LL55 4RF
Tel: 01286 870837
Contact: Fiona Brown Wilkinson
Established: 1985
Open: Usually 9am to 9pm
Craftworkers on site: 1
Crafts: Superb ceramic animal models
Prices: £1.50 to £250. Credit cards not accepted
Note: Close to mountain walks, Snowdonia, Padarn Park, Llanberis, Brynbras Castle and Dolbadarn Castle

CWM POTTERY
Trefor
Caernarfon
Gwynedd LL54 5NB
Tel: 01286 660545
Contact: R.E. Cheshire
Established: 1977
Open: 10am to 6pm. Advisable to telephone first
Craftworkers on site: 1
Crafts: Studio pottery – large individual shapes and domestic stoneware pottery
Prices: £4 to £180. Visa/Mastercard

INIGO JONES SLATEWORKS
Y Groeslon
Caernarfon
Gwynedd
LL54 7ST
Tel: 01286 830242
Fax: 01286 831247

Contact: D. Topliss
Established: 1861
Open: 9am to 4.30pm Monday to Friday
Craftworkers on site: 4
Crafts: Slate craft and historical letter cutting and calligraphy exhibition. Pottery, painting on slate
Prices: £1.50 to £100. Access/Visa
Note: Opportunity to try calligraphy and letter cutting on slate. Self guided tour around workshops comprises video and taped commentary

SEALED EARTH
'Fron' Allt Goch
Cwm y Glo
Caernarfon
Gwynedd
LL55 4DW
Tel: 01286 871170
Contact: Sue King
Established: 1991
Open: Please telephone for details
Craftworkers on site: 1
Crafts: Pottery – various types, sizes and prices, all smoke fired
Prices: £8 to £80. Credit cards not accepted
Note: Visitors welcome but preferably by appointment

CRAFT IN THE BAY
72 Bute Street
Cardiff Bay
South Glamorgan
CF1 6AX
Tel: 01222 484611
Fax: 01222 491136
Contact: Lauraine Voisey
Established: 1996
Open: 10am to 5pm Tuesday to Sunday
Craftworkers on site: 2
Crafts: Textiles, ceramics, wood turning
Prices: £1.50 to £2,000. Access/Visa/Mastercard/Switch/American Express etc.
Note: Craft demonstrations and work-shops, cafe, play area, free car/coach parking, disabled access (shop, gallery, all workshops)

ENJAY DESIGNS
17 Blenheim Road
Penylan
Cardiff
South Glamorgan
CF2 5DS
Tel: 01222 484776

Fax: 01222 456662
Contact: Emma Bettinson
Established: 1996
Open: Visitors by appointment only
Craftworkers on site: 2
Crafts/stock: Jewellery using semi-precious stones. Old, collectable and hand-made beads. Celtic silver jewellery
Prices: £5 to £100. Visa/Mastercard

MAKERS GALLERY
37 Penylan Road
Roath
Cardiff
South Glamorgan
CF2 3PG
Tel: 01222 472595
Contact: Helen Lush
Established: 1991
Open: 10am to 5.30pm. Closed Sundays and Mondays (open every day in December to Christmas)
Stock: Ceramics: porcelain, stoneware, earthenware, raku, sculptural – thrown and cast. Turned wood, stained glass, wooden toys, painted silk, embroidered pictures, woven tapestries and cushions. Patchwork and appliqué, watercolours and prints, jewellery, cards. All made in Wales to high standard
Prices: £1.50 to £600. Visa/Access/MasterCard
Note: Changing monthly exhibitions. Work by members of the Craft Co-operative on sale. 'Makers' in shop to advise and help. Commissions accepted

MUSEUM OF WELSH LIFE
St. Fagans
Cardiff
South Glamorgan
CF5 3XB
Tel: 01222 569441
Contact: Miranda Kennett
Open: 10am to 5pm daily (October to June). 10am to 6pm daily (July to September)
Crafts/stock: Miller, saddler, blacksmith, cooper, potter, baker and weavers. Pottery, metalwork, turned wood, glass
Prices: £5 to £100. Credit cards not accepted
Note: Craftsmen form part of the Museum of Welsh Life – a living history. Please telephone for admissions prices. Restaurant, teashop, craft demonstrations, fairs etc.

OLD LIBRARY CRAFT CENTRE SHOP
Makers Guild in Wales
Old Library, Trinity Street
Cardiff
South Glamorgan CF1 2BH
Tel: 01222 222584
Contact: Lauraine Voisey
Established: 1988
Open: 10am to 5.30pm. Closed Sundays.
Open every day in December up to
Christmas
Stock: All ceramics, textiles: woven,
knitted, patchwork, embroidered.
Jumpers, jackets, hats, scarves. Silver
jewellery, costume jewellery, turned
wood, wall hangings, cards.
Prices: £2 to £1,000. Amex/Switch/Visa/
Access/Mastercard
Note: Run by the Makers Guild in Wales.
Located in original Victorian listed
buildings, access via unique tiled
corridor.

ORIEL
The Arts Council of Wales Gallery
The Friary
Cardiff
South Glamorgan
CF1 4AA
Tel: 01222 399477
Fax: 01222 398500
Contact: Jenni Spencer Davis
Established: 1989
Open: Monday to Saturday 9am to
5.30pm
Stock: High quality domestic pottery,
studio ceramics, glass, wood, metal,
jewellery, textiles, furniture as well as
limited edition prints
Prices: From £10 upwards. Most major
credit cards are accepted
Note: City centre gallery with lively
range of contemporary craft from Wales
and the UK complemented by exhibitions

DAVID BEATTIE ETCHING STUDIO
Paisley Villa
Llandygwydd
Cardigan
Dyfed
SA43 2QT
Tel: 01239 682649
Contact: David Beattie
Established: 1988
Open: 10am to 4pm April to October.
Closed Wednesdays, Saturdays, Sundays
Craftworkers on site: 1

Crafts: Limited edition original prints
hand printed on premises by resident
artist. Greetings cards
Prices: £20 to £1,000. Visa/Mastercard/
Access/Eurocard
Note: Periodically visitors may see artist
printing and proofing new work

JOE & TRUDI FINCH
Tanygroes
Cardigan
Dyfed
SA43 2HR
Tel: 01239 810265
Contact: Joe Finch
Established: 1989
Open: Please telephone for details
Craftworkers on site: 2
Crafts: Pottery and paintings

JOHN & VICTORIA JEWELLERY
Ydlanddegwm
Llechryd
near **Cardigan**
Dyfed SA43 2PP
Tel: 01239 682653
Contact: John and Victoria Callen
Established: 1978
Open: Tuesday to Friday 10am to 6pm
May to September and December
Craftworkers on site: 2
Crafts: Handmade jewellery in silver and
gold
Pries: £3 to £200. Access/Visa/MasterCard
Note: Visitors can see jewellery being
made in workshop. Showroom adjoining.
Set in beautiful garden with outstanding
views of countryside and sea

ORIEL MYRDDIN
Church Lane
Carmarthen
Dyfed SA31 1LH
Tel: 01267 222 775
Contact: Sian Griffiths
Established: 1992
Open: Monday to Saturday 10.30am to
4.45 including Bank Holidays
Stock: Domestic pottery, studio ceramics,
wood, metal, jewellery, glass, toys,
textiles as well as prints
Prices: From under £20 up to £500.
Major credit cards accepted
Note: Changing display of contemporary
crafts by new and established makers in
Wales and UK. Exhibitions throughout
the year

ORIGIN DYFED
1 St. Mary's Street
Carmarthen
Dyfed SA31 1TN
Tel: 01267 220377
Contact: Narda Mantle
Established: 1996
Open: 10am to 5pm Tuesday to Saturday
Stock: Local crafts from pots to papier
mâché, candles to calligraphy and silk to
salt dough
Prices: 30p to £150. Credit cards not
accepted
Note: Situated in the heart of west Wales'
historic market town and features work of
some of our finest artisans. Gallery run by
Origin Dyfed, a non-profit making
community crafts co-operative with
another gallery in Newquay

MADY GERRARD
Grove Cottage, St. Arvans
near **Chepstow**
Gwent NP6 6EU
Tel: 01291 625764
Contact: Mady Gerrard
Established: 1985
Craftworkers on site: 4
Crafts: Women's high fashion collage
jackets
Prices: £75 to £145. Credit cards not
accepted

JIM HEATH WOODCARVER
2 Hand Terrace, Holyhead Road
Chirk
Clwyd LL14 5EU
Tel: 01691 772710
Contact: Jim Heath
Established: 1994
Open: Most days but please 'phone first
Craftworkers on site: 1
Crafts: Architectural and sculptural
woodcarvings
Prices: £5 upwards. No credit cards
Note: Visitors are welcome to watch
work in progress and are usually offered
tea/coffee and biscuits. Commissions
undertaken

THE SLATE WORKSHOP
Pont Hywel Mill, Llangolman
Clunderwen
Dyfed SA66 7XJ
Tel: 01994 419543
Fax: 01994 419543
Contact: Richard and Fran Boultbee

Established: 1985
Open: 9am to 5.30pm Monday to
Saturday and 9am to 4pm Sunday. Please
telephone in winter
Craftworkers on site: 2
Crafts: Lettering for housenames, plaques,
sundials and clocks; sculpture; lamp
bases, cheese boards, plant pots and
troughs, barometers, jewellery boxes
Prices: £10 to £2,000. Access/Visa/
Mastercard
Note: Demonstrations by appointment.
Located by eastern Cleddau River.
Commissions undertaken

CORWEN MANOR CRAFTS
8 London Road
Corwen
Clwyd LL21 0DR
Tel: 01490 413196
Contact: Alan Sayer and Sandra Sayer
Established: 1989
Open: Daily 10am to 5.30pm Easter to
November. Friday, Saturday, Sunday
10am to 4.30pm December to March
Stock: Ornate and novelty candles,
pottery, ornamental butterflies, slippers,
wrought iron (ornamental), slate and coal
figurines and tourist gifts
Speciality: Candlemaking and accessories
Prices: 99p to £50. No credit cards

Y GLASSBLOBBERY
Glanrafon Hall, Glanrafon
near **Corwen**
Clwyd LL21 0HA
Tel: 01490 460440
Fax: 01490 460247
Contact: Mrs. Wendy Pryce-Jones
Established: 1989
Craftworkers on site: 2
Open: Easter to end of October 10.30am
to 5.30pm. Closed Tuesdays. 7 days a
week July and August
Crafts/stock: High quality UK craft: glass,
pottery, wood, jewellery, cards, land-
scape painting and prints
Speciality: Glass sculptures
Prices: £1.99 to £1,000. All major
credit cards except Switch accepted
Note: Located in old Victorian hall 9
miles north of Bala on the A494. Regular
demonstrations of glassblobbing. Visiting
craftsmen give demonstrations and
evening talks. Coach car park, cafe and
garage. Member of Wales Craft Council.
Disabled access ground floor only

LITTLE RED DRAGON
1 Westgare, Eagle House
Cowbridge
South Glamorgan CF71 7AQ
Tel: 01446 775610
Contact: Monica Mahoney
Established: 1996
Open: 10am to 5pm Monday to Saturday
but half day Wednesday to 1pm
Stock: Welsh knitwear direct from the
Woollen Mills. Welsh language T-shirts,
ceramics and toys.
Prices: £1.50 to £160. Mastercard and
Visa accepted

MAKERS
74 Eastgate
Cowbridge
South Glamorgan
Tel: 01446 775280
Contact: Molly Curley
Established: 1984
Open: 10am to 5pm. Closed Sundays and
Wednesdays. December: open Monday
to Saturday till Christmas
Stock: Jewellery, watercolours, pottery,
painted silk, children's designer fashions,
iron work, turned wood, soft toys, painted
ceramic tiles, cards. Made locally to high
standard
Prices: £1.50 to £100. Visa/Access/
MasterCard
Note: Tiny 400-500 year old cottage-
type shop. Co-operative of ten 'makers'
who work in shop (a maker is always
present). Commissions accepted for
house portraits, commemorative plaques

CRICCIETH CRAFT CENTRE
The Square
Criccieth
Gwynedd LL52 0EP
Tel: 01766 522867
Fax: 01766 523522
Contact: Roy Williams
Established: 1965
Open: 9.30am to 5.30pm and 7.30pm to
10pm Easter to October
Stock: High quality slatecraft produced by
owner. House names and numbers,
clocks, barometers, pen stands etc. Craft
by local potters and mixed crafts
Speciality: Commemorative plaques/
memorials, marine barometers, clocks
Prices: £1.95 to £2,000. All major credit
cards accepted
Note: Welsh teashop adjacent

BRYN POTTERY
Eglwyswrw
Crymych
Dyfed
SA41 3SS
Tel: 01239 891608
Contact: Hilary Bassett
Established: 1987
Open: Summer: 10am to 6pm closed
Saturday and Monday. Winter: 10am to
5pm Wednesday, Thursday, Friday only
Craftworkers on site: 1
Crafts: Stoneware pottery
Prices: £5 to £100. Credit cards not
accepted
Note: Unique, well-designed, hand
made, beautiful, decorative, functional
stoneware

WEST WALES ART CENTRE
Incorporating Talking Point
Castle Hill
16 West Street
Fishguard
Pembrokeshire
SA65 9AE
Tel: 01348 873867
Contact: Myles Pepper
Established: 1983
Open: Monday to Saturday 9am to
5.30pm. Sunday by appointment
Stock: Domestic pottery, studio ceramics,
jewellery as well as original prints and
paintings
Prices: £20 to £1000. Most credit cards
accepted
Note: Contemporary applied arts and fine
art plus lectures and music recitals

BWTHYN CRAFTS
Oxwich
Gower
near Swansea
West Glamorgan
SA3 1LS
Tel: 01792 390686
Contact: Nelian Cook
Established: 1985
Open: Easter to October. Telephone for
details of out of season times
Stock: Ceramics, prints, soft toys,
jewellery, love spoons, Dylan Thomas
mural, celtic slate and terracotta
Prices: £1 to £25. Visa/MasterCard
Note: All crafts made in Wales. Local
interest books and maps and ice cream
also available

PEWTER STUDIO
Kerrigwyn
Rectory Road, Llangwm
Haverfordwest
Dyfed SA62 4JA
Tel: 01437 891318
Contact: Paul Morafon
Established: 1987
Open: 9am to 5pm
Stock: Hand-embossed pewter mirrors,
photoframes and jewellery
Note: Located in beautiful scenery

TREAMLOD TILERY
Ambleston
Haverfordwest
Dyfed SA62 5OQ
Tel: 01437 741541
Contact: Chris Cox
Established: 1992
Open: 10am to 6pm daily except Sunday
and Thursday
Craftworkers on site: 1
Crafts: Medieval and Victorian floor tiles
Note: Demonstrations of medieval and
Victorian tile manufacture. Display of
original tiles. Handmade tiles for sale and
commissions undertaken

ORIEL LLYNFAES
Llynfaes Uchaf
Lynfaes, Tynlon
near **Holyhead**
Anglesey LL65 3BJ
Tel: 01407 720434
Contact: Allen
Established: 1995
Open: 9am to 5pm Friday to Sunday
Craftworkers on site: 1
Crafts: Sculpture and painting
Prices: Various. Credit cards not accepted
Note: Organic display gardens – flowers,
fruit and vegetables, trees, shrubs.
Sculpture park and art gallery. Coffee
available. Ample parking and disabled
access

**BARBARA WARD (HAND MADE
CERAMICS)**
Church Farm, Revnalton
Kilgetty
Dyfed SA68 0PG
Tel: 01834 891211
Contact: Garrick Ward
Established: 1969
Open: During summer. Please telephone
for details

Craftworkers on site: 3
Crafts: Cast creamware, spongeware,
hand painted items. Tiles, plates
whiteware
Prices: £8 to £80. Visa card accepted
Note: Located in old city wall Tenby.
Demonstrations held. Commissions
undertaken. Small runs of specialised
orders for pubs, rotary clubs, etc.

CAER SIDI CRAFT
Station Crescent
Llandrindod Wells
Powys LD1 6AS
Tel: 01597 825044
Contact: Richard and Heather Grosvenor
Established: 1990
Open: 9.30am to 5.30pm Monday to
Saturday
Stock: Celtic jewellery, arts and crafts
from the celtic countries. High quality
gifts and souvenirs
Speciality: Celtic jewellery
Prices: £2 to £150. All major credit cards
accepted

PATSY BESWICK
Guidfa House, Crossgates
Llandrindod Wells
Powys LD1 6RF
Tel: 01597 851241
Contact: Patsy Beswick
Craftworkers on site: 1
Open: Please telephone for details
Crafts: Hand spun wool, fleece rugs,
hand knitted goods
Prices: £5 to £200. Credit cards not
accepted
Note: Workshops by arrangement or
individual tuition available

PORTICUS
1 Middleton Street
Llandrindod Wells
Powys LD1 5ET
Tel: 01597 823989
Contact: Rosemary Studman
Established: 1988
Open: Monday to Saturday 10am to 5pm.
Early closing Wednesday
Stock: Domestic pottery, studio ceramics,
glass, wood, jewellery, textiles and knitwear
Prices: From under £5 to over £200. Most
major credit cards accepted
Note: Wide selection of contemporary
craft. Changing exhibitions throughout
the year. Set in Victorian surroundings

ORIEL MOSTYN CRAFT & DESIGN SHOP
12 Vaughan Street
Llandudno
Gwynedd
LL30 1AB
Tel: 01492 879201/870875
Fax: 01492 878869
Contact: Sue Evans
Established: 1982
Open: Monday to Saturday 10am to 5pm, closed 1pm to 1.30pm. Closed Sundays and Christmas
Stock: Domestic pottery, studio ceramics, glass, wood, jewellery, textiles, toys,
Prices: From around £5 up to £300. Most major credit cards accepted
Note: Situated in Wales' premier art gallery. Constantly changing display and monthly exhibitions

FANTASY
Unit 2, The Museum of the Welsh Woollen Industry
Drefach Felindre
Llandysul
Dyfed SA44 5UP
Tel: 01559 371299
Fax: 01559 371299
Contact: Helene and Paul
Established: 1994
Open: May to October 9.30am to 4pm weekdays, 11am to 4pm Saturdays. October to April 10am to 4pm weekdays only
Craftworkers on site: 2
Crafts/stock: Candlemaking and essential oil bottling and blending. Candles, incense, aromatherapy related items
Speciality: Naturally scented candles
Prices: 11p to £25. Visa/Mastercard
Note: Museum site with other crafts workshops. Car park, picnic area and teashop

MAKEPEACE CABINETMAKING
Derw Mill
Pentrecwrt
Llandysul
Dyfed
SA44 5DB
Tel: 01559 362322
Fax: 01559 362322
Contact: Sarah Makepeace
Established: 1970
Open: Weekdays 9am to 5pm. Saturday 10am to 2pm

Craftworkers on site: 10
Crafts: Bespoke fitted and free-standing furniture made to the highest standards
Prices: £50 plus
Note: Picnic tables by stream and large car park. Not suitable for disabled

SUN CERAMICS
National Museum of Wales
Drefach Velindre
Llandysul
Dyfed SA44 5UP
Tel: 01559 371206
Fax: 01559 371206
Contact: Jez Waller and Lesley Davies
Established: 1988
Open: 9am to 5pm
Craftworkers on site: 5
Crafts: Garden terracotta, pots, plaques, fountains etc.
Prices: £1 to £120. Credit cards not accepted
Note: Other craft workshops, woollen museum, cafe, parking for coaches and cars, water wheel. Member of Wales Craft Council, Origin Dyfed

LLANGEDWYN MILL
Llangedwyn
near Oswestry
Powys SY10 9LD
Tel: 01691 780528
Fax: 01691 780601
Contact: Mrs. Maureen Wilde
Established: 1980
Open: 10am to 5pm Monday to Friday all year round. Weekends by appointment
Craftworkers on site: 3
Crafts: Stained glass studio (Mirage Glass 01691 780618). Jewellery/clock restoration (01691 780642). Ceramics (01691 780140)
Prices: £5 to £1,000. Visa/Mastercard/Euro/Switch/JCB (Glass workshop only)
Note: Attractive riverside site, free car park, toilets. Disabled access

MIRAGE GLASS
Units 9-11, Llangedwyn Mill
Llangedwyn
near Oswestry
Powys SY10 9LD
Tel: 01691 780 618
Fax: 01691 780618 (10-5 only)
Contact: Wolfe and Johan van Brussel
Established: 1978
Open: 10am to 5pm Monday to Friday

all year. Weekends by appointment
Craftworkers on site: 2
Crafts: Stained glass – architectural, also decorative – lampshades, mirrors, boxes, screens, small gifts, jewellery
Prices: £3.50 to £500. Visa/MasterCard/Switch/JCB/Access/Eurocard
Note: Most of our work is architectural stained glass to commission, but visitors welcome. Occasional day and weekend courses held. Potter and silversmith also on site. Free parking, toilets, picnic area by river, tearoom in summer

PERIOD DOLLS WALES
Cwm Bach
Llanddeiniol
Llanrhystud
Dyfed SY23 5AN
Tel: 01974 202684
Fax: 01974 202684
Contact: Marjorie Hill
Established: 1992
Open: Most days but please phone first
Craftworkers on site: 2
Crafts: Porcelain collector dolls, Welsh and celtic costumes. Exclusive Welsh costume prints, mounted prints, greeting cards, postcards from the archives of the National Library of Wales
Prices: £5 to £150+. Credit cards not accepted
Note: Member of Wales Craft Council

TREFRIW WOOLLEN MILLS LTD
Trefriw
near **Llanrwst**
Gwynedd LL27 0NQ
Tel: 01492 640462
Fax: 01492 640462
Contact: Elaine Williams
Established: 1859
Open: Monday to Friday 10am to 5pm
Craftworkers on site: 6
Crafts: Traditional Welsh bedspreads, knittings wools, mohair rugs, pottery, knitwear, outdoorwear
Prices: £1.25 to £100. Mastercard/Visa/Switch/American Express
Note: Tearoom, weavers' garden with specimen plants used in textile production – flax, woad, weld, madder, soapwort, teazles etc. Handspinning, "Try weaving yourself" display of small looms (Spring Bank Holiday to end September). Weaving and hydro-electric turbines can be viewed all year

CHRISTINE GITTINS STUDIO POTTER
Model House Craft & Design Centre
Bull Ring
Llantrisant
Mid-Glamorgan
CF72 8EB
Tel: 01443 237758
Fax: 01443 224718
Contact: Christine Gittins
Established: 1994
Open: Please telephone for details
Crafts: Ceramics
Prices: £8 to £100. Visa/MasterCard
Note: Situated in craft centre. Professional member of the Craft Potters Association and Makers Guild in Wales

HILL TOP STUDIO
56 High Street
Old Town
Llantrisant
Mid-Glamorgan
CF7 8BN
Tel: 01443 229015
Contact: Rachael Evans
Open: 10am to 5pm Thursday, Friday, Saturday and Sunday
Stock: Dried and silk flowers, pottery, prints, paintings, country pine furniture, books, unusual crafts, candles
Prices: 50p+. Most major credit cards accepted
Note: Butchers Arms Gallery and Coffee Shop, Traditional Toys shop and Model House Crafts & Design Centre nearby

MODEL HOUSE CRAFT & DESIGN CENTRE
Bull Ring
Llantrisant
Mid-Glamorgan
CF7 8EB
Tel: 01443 237758
Fax: 01443 224718
Contact: Jenny Rolfe
Established: 1989
Open: 1 May to 22 December: Tuesday to Sunday 10am to 5pm. 2 January to 30 April: Wednesday to Sunday
Stock: Ceramics, glass, jewellery, turned and carved wood, black iron
Prices: £5 to £500. Visa/Access
Note: Workshop and exhibition programme. Permanent exhibition from Royal Mint Exhibition of Artefacts and History of Llantrisant Castle. Talks by resident craftspeople arranged for groups

CAMBRIAN WOOLLEN MILL
Llanwrtyd Wells
Powys LD5 4SD
Tel: 01591 610211
Fax: 01591 610399
Contact: Susan Beetlestone
Established: 1820
Open: 9.30am to 4.30pm
Crafts: Wool processing/weaving
Prices: £1 to £100. Visa/Access/Amex/ Switch
Note: Tearoom and nature trail and tour

RICHARD WITHERS
1 Corris Craft Centre
Corris
Machynlleth
Powys SY20 9RF
Tel: 01654 761249
Contact: Richard Withers
Established: 1982
Open: 10am to 6pm summer season 7 days. 10am to 4pm winter 5 days. Closed between Christmas and 1st week of January
Crafts/stock: Kitchenware, decorative and practical wooden items, artistic pyrography. Framed poems and prose in calligraphy all hand penned
Speciality: Artistic pyrography
Prices: £1.25 to £1,500. Credit cards accepted
Note: This is a working craft centre and visitors may see craftworkers in action. T.I.C. and restaurant/café on site and gardens with picnic area. Adventure playground adjacent to centre and a theme park – King Arthur's Labyrinth – in adjoining quarry

AFONWEN CRAFT AND ANTIQUE CENTRE
Afonwen
near **Mold**
Clwyd CH7 5UB
Tel: 01352 720965
Fax: 01352 720346
Contact: David and Robert Monshin
Established: 1992
Open: Tuesday to Sunday all year. Closed Mondays. Open Bank Holidays. 9.30am to 5.30pm
Stock: Quality crafts, wool, cotton sweaters, cardigans etc. Craft kits for all ages. Antiques, pine etc. Please phone for details of exhibitions and working demonstrations

Prices: 75p to £2,350
Note: Originally Victorian Mill. Restaurant and six shops. Centre is sign posted from main road. Large car park. No entrance fee

TRI THY CRAFT & NEEDLEWORK CENTRE
Coed Talon
near **Mold**
Clwyd CH7 4TU
Tel: 01352 771359
Fax: 01352 771881
Contact: Jane Restall
Established: 1980
Open: Tuesday to Saturday 10.30am to 5pm. Closed Sunday and Monday all year
Stock: Needlework kits/accessories, woodturning, china painting, canalware, jewellery
Speciality: Needlecrafts
Prices: 99p to £100. Visa/MasterCard
Note: Tearoom, workshop, gallery. Exhibitions held. Signposted off A5104 Chester/Corwen Road

WENDI'S ART & CRAFT STUDIO
Loggerhead Country Park
Loggerheads
near **Mold**
Clwyd CH7 5LH
Tel: 01352 810458
Contact: Wendi Williams
Established: 1990
Open: 12 noon to 5pm daily including Saturday and Sunday
Stock: Watercolour paintings, painted glass, painted silk, candles, jewellery, wood turning (commissions undertaken)
Prices: 25p to £200. Cheques with Bankers' cards accepted
Note: Artist at work in shop. Countryside Park location with nature and industrial trails, information centre and cafe. Riverside and public house opposite park.

COUNTRY WORKS GALLERY
Broad Street
Montgomery
Powys SY15 6PH
Tel: 01686 668866
Contact: Richard & Clare Halstead
Open: April to Christmas, Tuesday to Saturday 10am to 5.30pm. Sunday 2pm to 5.30pm. Mid January to March, Wednesday to Saturday 10am to 5.30pm
Stock: Domestic pottery, studio ceramics,

glass, wood, metal, jewellery, textiles, knitwear and some unframed prints
Prices: From under £10 to over £6000. Most major cards accepted
Note: Monthly exhibition held of both established and new makers of paintings and sculpture

THE GOLDEN SHEAF GALLERY
High Street
Narberth
Dyfed SA67 7AR
Tel: 01834 860 407
Contact: Suzanne
Established: 1993
Open: Monday to Saturday 9.30am to 5.30pm. Extended opening during summer and Bank Holidays
Stock: Domestic pottery, studio ceramics, wood, metal, jewellery, textiles, knitwear, glass, toys, hats as well as original paintings
Prices: From under £5 up to £1500. Most major credit cards accepted
Note: Handsome Georgian building in charming town. Comprehensive selection of exciting contemporary art & craft

GLASSCRAFT
16 Pentre-Poeth Road, Bassaleg
Newport
Gwent NP1 9LL
Tel: 01633 895786
Fax: 01633 895786
Contact: Lionel and Judy Johnson
Established: 1988
Open: Most days but please 'phone first
Crafts: Glass of all descriptions
Speciality: Club and association trophies
Prices: £1.99 to £50. Credit cards not accepted
Note: Glass engraving while you wait

CREFFT QUAY CRAFTS
3rd Floor Slipway Buildings
The Old Slipway
Newquay
Dyfed SA45 9PS
Tel: 01545 561067
Contact: Narda Mantle
Established: 1993
Open: 10am to 6pm 7 days a week peak season. 11am to 4pm 4 days a week at other times
Craftworkers on site: 1
Stock: This non-profit making co-operative gallery run by local makers shows a vast range of locally made arts,

crafts, cards and prints
Prices: 30p to £200
Note: Occasional craft demonstrations held. Wonderful view overlooking Newquay Bay. Sister gallery of equal distinction in Carmarthen. Newquay is built on a steep hill so disabled access difficult but possible

OLD MILL CRAFT SHOP/POTTERY
Flour Shed Gallery
Snowdon Mill, Snowdon Street
Porthmadog
Gwynedd LL49 9DF
Tel: 01766 510912
Fax: 01766 510913
Contact: Jane Williams
Established: 1975
Open: 9.30am to 4.30pm
Craftworkers on site: 10
Crafts/stock: Pottery, resin cast cottages, candles, gift and craft items
Prices: 50p to £1,000
Note: Workshop visitor centre, hands on crafts activities, courses held, tearoom

OVER THE RAINBOW
74 High Street
Pwllheli
Gwynedd LL53 5RR
Tel: 01758 612989
Contact: Linda and Dewi Jones
Established: 1985
Open: 10am to 5pm Monday to Saturday. Closed Thursdays, Christmas to Easter
Stock: Handmade crafts, activity kits, cards, prints and original paintings, art materials, jewellery
Speciality: British crafts, prints and paintings by local artists
Prices: £1 to £300. Visa/Mastercard/ American Express

THE GALLERY
Ruthin Craft Centre, Park Road
Ruthin
Clwyd LL15 1BB
Tel: 01824 704774
Contact: Jane Gerrard and Philip Hughes
Established: 1982
Open: Summer 10am to 5.30pm Monday to Sunday. Winter 10am to 5pm Monday to Saturday, 12 noon to 5pm Sunday
Stock: Contemporary applied art including jewellery, ceramics, glass, textiles, automata etc.
Prices: £5 to £1,000. Visa/Mastercard

POTTER'S TROTTERS
12 Saunders Way
Derwen Fawr
Swansea
West Glamorgan SA2 8AY
Tel: 01792 204301
Contact: Brian Harrison
Established: 1993
Open: Please telephone for details
Craftworkers on site: 1
Crafts: Rocking horses and terracotta
Prices: £35 (pots) to £1600 (horses)
Note: Visitors welcome by arrangement

"IN THE PINK"
24 Commercial Street
Tredegar
Gwent NP2 3DH
Tel: 01495 711044
Contact: Claire Smith
Established: 1995
Open: 9.30am to 5pm. Early closing
Thursday 1.30pm
Stock: China and ceramics. Greeting
cards, wooden stools and chairs. Wooden
gifts, hand turned coal figures. Hand
painted clocks, corn dollies. Many items
can be personalised or made to order
Prices: 50p to £50. Credit cards not
accepted
Note: All items made in Wales mostly in
Gwent. Also two china and greeting cards
workshops

THE WELSH GOLD CENTRE
Main Sqare
Tregaron
Dyfed SY25 6JL
Tel: 01974 298415
Fax: 01974 298690
Contact: Joelle Moggan
Established: 1971
Open: 9.30am to 5.30pm Monday to
Saturday
Craftworkers on site: 2
Crafts: Gold and silver celtic design
jewellery
Prices: 99p to £700. Visa/Mastercard/
Amex/Switch and most cards
Note: Home of the famous Rhiannon
jewellery and designs in Welsh gold

RHOSLAN WOODS
The Old Mill Workshop
Kilkewydd Mill, Forden
Welshpool
Powys SY21 8RT

Tel: 01938 553383
Contact: Lorraine Lloyd
Established: 1995
Open: 10am to 4pm
Craftworkers on site: 2
Crafts: Welsh hardwood gifts, souvenirs
and toys
Prices: 50p to £100. Credit cards not
accepted.
Note: Craft demonstrations held on
various woodworking skills. Display on
promotion of local hardwoods; different
types and uses. Enjoyable and educa-
tional

HAFOD HILL POTTERY
Hafod Hill
Llanboidy
Whitland
Dyfed SA34 0ER
Tel: 01994 448361
Contact: Roger Brann and Christine
McCole
Established: 1980
Open: Monday to Friday 9am to 5pm all
year
Craftworkers on site: 2
Crafts: Domestic stoneware, hand
thrown and wood fired
Prices: £1.50 to £30. Credit cards not
accepted
Note: Working pottery open to visitors
who can see various processes. Quiet
rural location, nearby walks etc.

STUDIO IN THE CHURCH
near Login
Whitland
Dyfed
SA34 0XA
Tel: 01437 563676
Contact: Alan Hemmings
Established: 1978
Open: 10am to 5pm Monday to Friday.
Please telephone for details in winter
Craftworkers on site: 2
Crafts/stock: Handweaving cloaks,
ruanas, hats, fashion accessories in
mohair, pottery, brooches, selected crafts
and Pembrokeshire paintings
Prices: £3 to £150. Visa/Mastercard/
Eurocard
Note: Demonstrations held. Mini coffee
shop. Studio/gallery in restored ancient
church in Preseli foothills near
Carmarthen/Pembrokeshire border.
Disabled access - wheelchairs welcomed

APPLIQUÉ EMBROIDERY
The Plassey Craft Centre
Eyton
near **Wrexham**
Clwyd
LL11 0SP
Tel: 01978 780916
Contact: John and Jennifer
Established: 1987
Open: 10am to 5pm 6 days a week all year. Closed Thursdays
Crafts/stock: Appliqué cushions, cards, tea cosies, pin cushions, needlework accessories and kits
Prices: £1.95 to £30. Credit cards not accepted
Note: Part of lovely craft centre complex with tea garden, coffee shop, bistro, nature walk, lake with wild fowl, golf course and touring caravan park etc. Free admission and free parking

THE PLASSEY CRAFT CENTRE
Eyton
Wrexham
Clwyd
LL13 0SP
Tel: 01978 780277
Fax: 01978 780019
Contact: Mrs. Della Brookshaw
Open: All year 10.30am to 5pm. Closed some Mondays – please telephone
Craftworkers on site: 12
Crafts: Dried flower arrangements, appliqué embroidery – cushions etc. Paperweights, ceramics, picture framing, woodturning – bowls etc. Aromatherapy oils and beauty preparations
Prices: £1 to £100. Credit cards not accepted
Note: Woodturning demonstrations held. Hobby ceramics classes, garden centre, beauty salon, hair studio, coffee shop with tea garden. Disabled access. Coaches by arrangement. Touring caravan park adjoins craft centre with many amenities

REBECCA OSBORNE STUDIO
Foxbrush Gardens
Portdinorwic
Y Felinheli
Gwynedd
LL56 4JZ
Tel: 01248 670463
Contact: Rebecca and Jenny Osborne
Established: 1992

Open: By prior appointment only
Craftworkers on site: 2
Crafts: 3D miniature scenes and hand grown plants
Prices: 50p to £300. Credit cards not accepted
Note: Award winning 3D scenes exhibited in galleries throughout Wales. Three acre country garden open to the public, cottage museum, plants for sale, teas served in indoor jungle. Very accessible for wheelchairs

THE STEVEN JONES GALLERY
22 Bangor Street
Y Felinheli
Gwynedd
LL56 4JD
Tel: 01248 671459/671209
Contact: Steven Jones
Established: 1990
Open: Monday to Saturday 9.30am to 5.30pm
Craftworkers on site: 1
Crafts: Paintings, limited edition prints and greeting cards
Prices: £1.20 to £800. Credit cards not accepted
Note: Work undertaken for advertising agencies, publishers, magazines etc. Disabled access

The outlets of various types that we have listed so far in this book are an important way of finding crafts and making contact with craft makers but not the only way. The UK has an established circuit of craft fairs; indeed, there are scores of them taking place each weekend. Our publication for professional craftworkers *(The Craftworker's Year Book)* lists thousands of such events and drawing on this knowledge we provide below a list of organisers who, we believe, put on interesting, varied events of above average quality.

A phone call to any organiser with a request for an up-to-date list of their events should provide you with the key to accessing hundreds more craftworkers, many of whom do not sell direct from their studios and have not therefore been listed elsewhere in this book

In respect of that vexed question of quality we have provided a coding to indicate as far as is possible the type of selection policy operated by each firm. Where, in an entry, the letter **A**

appears against Selection Policy, the firm's declared regime is "Very, very strict – probably only the highest quality of design, display and skill allowed, in many cases from graduate designers/artists/craftworkers". For **B** – "Highly considered – all craft, makers only and only allowed after full examination of work and/or photographs" while the letter **C** is "Liberal– mostly craft, favouring maker exhibitors, but only gentle or no sanction exercised as to design ability or skill level". The last option, **D** is "Highly flexible – bought-in goods permitted, no sanction as to quality exercised".

We must emphasise that while every effort has been made to ensure the accuracy of this information (and indeed all data in this book) interpretation of what constitutes "quality" and what does not is a matter of opinion and we cannot except any liability financial or otherwise for our opinion expressed or for error or omission. Those entries appearing all in bold type are advertisers or organisers that have paid for that service.

3D2D CRAFT & DESIGN FAIRS
Unit 3 – Albion Business Centre
78 Albion Road
Edinburgh
Midlothian EH7 5QZ
Tel: 0131-661 6600
Fax: 0131-661 0012
Contact: Kathy Mills/Susan Webster
Number of Events: 60
Selection policy: B
Major organiser of events in Scotland

ANGELA WOODHOUSE ROAD SHOW
37 Awsworth Lane
Cossall
Nottinghamshire NG16 2SA
Tel: 0115 944 4489
Contact: Angela Woodhouse
Number of Events: 10
Selction policy: C

ANN BESWICK
Guidfa House
Crossgates
Llandrindod Wells
Powys LD1 6RF
Tel and Fax: 01597 851875
Contact: Anne Beswick
Number of Events: 4

Selection Policy: B
Established events aimed at discerning
buyers showing the best of British Crafts

ART IN ACTION
96 Sedlescombe Road
Fulham
London SW6 1RB
Tel: 0171-381 3192
Fax: 0171-381 0605
Contact: Bernard Saunders
Number of Events: 1
Selection policy: B
Annual event first run in 1976 which constitutes one of the largest shows of craft demonstrators in UK. Held at Waterperry House, Oxfordshire in mid July.

BRIDLINGTON MODEL BOAT SOCIETY
46 Manorfield Avenue
Driffield
East Yorkshire YO25 7HP
Tel: 01377 252550
Fax: 01377 252550
Contact: John Foster
Number of Events: 1
Selection policy: C
Largest model boat, hobbies and crafts festival in the North East

BUCKINGHAM PUBLICITY
11 Old High Street
Old Portsmouth
Hampshire   PO1 2LP
Tel: 01705 871000
Fax: 01705 297661
Contact: Christine Roberts
Number of Events: 25
Selection policy: B
Exhibitions and shows in stately homes
etc. all over England

CFL CRAFT SHOWS
Four Lanes End Farm
Hulland Ward, Ashbourne
Derbyshire DE6 3EJ
Tel: 01335 370078
Mobile: 0976 368989
Contact: Lee or Eleanor Israel
Number of Events: 20
Selection policy: B
Shows in Derbyshire, Warwickshire,
Cheshire, Northumberland, Yorkshire and
Cumbria

CLASS 6 PROMOTIONS
8 Ladycroft
Cubbington
Leamington Spa
Warwickshire CV32 7NH
Tel: 01926 426992
Contact: Rob Morgan
Number of Events: about 10
Selection Policy: B
Organise events at two superb venues in
Warwickshire

COTTAGE INDUSTRY CRAFT SHOWS
Huntsman's Cottage
Kennel Lane
Windlesham
Surrey GU20 6JQ
Tel: 01276 472132
Fax: 01276 479255
Contact: Brenda Gosling
Number of Events: 8-10
Selection policy: A/B – plus gift and
design work. 20th year organising shows
in Berkshire and Surrey

COUNTRY COTTAGE CRAFTS
Trunch Hill
Denton
Harleston
Norfolk IP20 0AE
Tel: 01986 788757
Fax: 01986 788856

Contact: David Hicks
Number of Events: 6
Selection policy: A/B
Events in East Anglia

CRAFT LINCS
PO Box 49, Grantham
Lincolnshire NG32 3SW
Tel: 01400 230043
Fax: 01400 230574
Contact: Mrs. Tracy Evans
Number of Events: 20
Selection policy: A/B
Crafts at major events in the East of
England area

CRAFT*FOLK
PO Box 26
Dinas Powys
South Glamorgan CF64 4YR
Tel: 01222 514732
Fax: 01222 514732
Contact: Howard Potter
Number of Events: 6+
Selection policy: B
Special craft events, mainly in Cardiff

CRAFTS COUNCIL
44a Pentonville Road
Islington, London N1 9BY
Tel: 0171-278 7700
Fax: 0171-837 6891
Contact: John Whiddett
Number of Events: 1
Selection policy: Juried event
Organiser of the renowned Chelsea Crafts
Fair

CRAFTS IN ACTION
53 Windmill Hill, Enfield
Middlesex EN2 7AE
Tel: 0181-366 3153
Fax: 0181-366 6844
Contact: Thalia West
Number of Events: 30
Selection policy: B
Craft marquees at major shows all around
the country

CRAFTWORKS
17-19 Linenhall Street
Belfast BT2 8AA
Tel: 01232 236334
Fax: 01232 236335
Contact: Hazel McAnally
Heavily involved in craft events and
promotion in Ulster

CRUNCHY FROG CRAFTS
PO Box 119, Newark
Nottinghamshire NG24 1QB
Tel: 01636 704382
Contact: Heather Smith
Number of Events: 20
Selection policy: B/C
Craft and 'Craft A Gift' events in Notting-
hamshire, Lincolnshire and Leicestershire

DALESWAY FESTIVALS LTD
"Ulvik"
Coneygarth Lane
Haxey
Doncaster
South Yorkshire
DN9 2JG
Tel: 01427 753040
Fax: 01427 753513
Contact: Graham Markwell
Number of Events: 10
Selection policy: A
Prestigious festivals of craft, fashion and
design at quality venues.

DULWICH CRAFT FAIRS
25 Tewkesbury Avenue
Forest Hill
London SE23 3DG
Tel: 0181-291 0024
Contact: Nicholas Keogh
Number of Events: 3
Selection policy: A
Small but premier rated events in
Dulwich Village in London

EAST KENT FAIRS
134/135 London Road, Dover
Kent CT17 0TG
Tel: 01304 201644 (closed Wednesdays)
Contact: John Payne
Number of Events: 20
One of the most active organisers around
the Cinque Ports region

EASTERN EVENTS LTD
Diggins Farm House
Buxton Road, Aylsham
Norfolk NR11 6UB
Tel: 01263 734711
Fax: 01263 735134
Contact: John Wootton/Janet Harker
DesRCA, MCSD
Number of Events: 10
Selection Policy: A/B
Shows at National Trust properties and
similar in England and Wales

FOCAL POINT
(FP Exhibition Management)
76 Main Road
Long Bennington
near Newark
Nottinghamshire
NG23 5DJ
Tel: 01400 281937
Fax: 01400 282051
Number of Events: 2
Selection Policy: A
High quality design-led exhibitions
selling to the gifts and crafts trade - and
public access may not be possible

FOUR SEASONS (EVENTS) LTD
23 Brockenhurst Road
South Ascot
Berkshire
SL5 9DJ
Tel: 01344 874787
Fax: 01344 874673
Contact: Ann Steele
Number of Events: about 20
Selection Policy: B
Organiser of major themed craft fairs at
prestigious locations in the South of
England

HAGLEY CRAFTS (96)
8 Fiery Hill Road
Barnt Green
Birmingham
West Midlands
B45 8LF
Tel: 0121-445 3967
Contact: Joyce Towers
Number of Events: 7
Selection policy: C
Hall/centre based events in West
Midlands, Warwickshire and Worcester-
shire

HOBBY-HORSE DESIGN & CRAFT FAIRS
PO Box 3751
Solihull
West Midlands
B91 3QF
Tel: 0121-711 4728
Fax: 0121-709 1012
Contact: Mrs Hilary Farnham
Number of Events: 20
Selection policy: A/B
Long established organiser of craft shows
and thematic craft events in and around
the West Midlands

JEAN WELCH SHOWS
West House
Crag Lane
Knaresborough
North Yorkshire
HG5 8EE
Tel: 01423 867114
Fax: 01423 867114
Contact: Jean Welch
Number of Events: 20
Selection policy: B
A well known organiser of craft marquees
and festivals at well known venues and
events in the North

KELD CRAFT FAIRS
Hilltop
Keld
Richmond
North Yorkshire
DL11 6LP
Tel: 01748 886260
Contact: John Morgan
Number of Events: 65
Selection policy: B
A major organiser of smaller events in the
rural North

KEVIN MURPHY CRAFT FAIRS
Chapel Farm
St. John's Chapel
Eastacombe
Barnstaple
Devon
EX31 3PB
Tel: 01271 43160
Mobile: 0374 904027
Contact: Kevin Murphy
Number of Events: 50
Selection policy: B
Lots of events in popular South West
locations

LFT EXHIBITIONS
Unit 10
Fernley Green Road
Knottingley
West Yorkshire
WF11 8DH
Tel: 01977 673777
Fax: 01977 673098
Contact: Michelle Jempson
Number of Events: 6
Selection policy: D
A relatively new organiser of craft
marquees at major shows in the East and
North of England

LIVING CRAFTS
1 The Chowns
West Common
Harpenden
Hertfordshire
AL5 2BN
Tel: 01705 817291/01582 761235
Contact: Jean Younger or Robin
Younger
Number of Events: 1
Selection policy: A
This event at Hatfield House is
Europe's superb and longest established
craft event. Held in May it presents a
wide range of crafts of very high
quality and certainly is not one to miss

LONDON BOROUGH OF ENFIELD
Countryside and Tourist Team
Civic Centre, Silver Street
Enfield
Middlesex
EN1 3XJ
Tel: 0181-982 7043
Fax: 0181-982 5450
Contact: David Smith
Number of Events: 1
Crafts at the Steam & Country Show,
Cockfosters which takes place over the
August Bank Holiday weekend

MARATHON EVENT MANAGEMENT
53 Windmill Hill
Enfield
Middlesex
EN2 7AE
Tel: 0181-366 3153
Fax: 0181-366 6844
Contact: Amanda Coomber/Mike Nugent
Number of Events: 3
Selection policy: A/B
Organiser of the British Craft Trade Fair in
Harrogate and retail events at Alexandra
Palace, North London

MARY HOLLAND CRAFT FAIRS LTD
PO Box 43
Abingdon
Oxfordshire
OX14 2EX
Tel: 01235 521873
Fax: 01235 521873
Contact: Pauline Burren
Number of Events: 3
Long established organiser of Prestige
shows at Mentmore, Petworth and
Abingdon

MGA FAIRS
PO Box 282
Overstone
Northampton
NN6 0SD
Tel: 01604 642185
Fax: 01604 642185
Contact: Marion or George Aldis
Number of Events: 7
Selection policy: A
Well established events in Leicestershire,
Northamptonshire, Warwickshire and
Cambridgeshire showing a wide range of
beautiful hand made good by British
crafts people

MIDLAND CRAFT FAIRS
Tang Beck House
Swincliffe Road
High Birstwith
near Harrogate
North Yorkshire
HG3 2JP
Tel: 01423 770925
Fax: 01423 770925
Contact: Robin Boardman
Number of Events: 8
Selection policy: B
Quality events in the Heart of England

NATIONAL CRAFTS FAIR
National House
28 Grosvenor Road
Richmond
Surrey
TW10 6PB
Tel: 0181-940 4608
Fax: 0181-891 0115
Contact: Anthony James
Number of Events: 3
Selection policy: A
The very highest quality - no imports or
bought in goods - very select fairs mainly
in the Channel Islands and Isle of Wight

NICKY MCGARRY
31 Seahill Road
Craigavad
Holywood
County Down
BT18 0DJ
Tel: 01232 422274
Fax: 01232 422274
Contact: Nicky McGarry
Number of Events: 12
Selection policy: B
A variety of venues in Ulster

OAK CRAFT FAIRS
7 Sandstone Avenue, Walton
Chesterfield
Derbyshire S42 7NS
Tel: 01246 569698
Contact: Chris Warriner
Number of Events: 15
Selection policy: C
East Midlands and South Yorkshire events

ORCHARD EVENTS LTD
1 Newton Grove
London W4 1LB
Tel: 0181-742 2020
Fax: 0181-995 0977
Contact: Sue Joy
Number of Events: 2
Selection policy: B
Country Style at the NEC, Birmingham

PHOENIX CRAFT FAIRS
59 Bankfield Road
Liverpool
Merseyside L13 0BD
Tel: 0151-228 1492
Fax: 0151-280 5190
Contact: Nikki or Brian Andrews
Number of Events: 34
Selection policy: B
Craft marquees in England, Wales and
Scotland; also shopping centres

PUBLICITY EXHIBITIONS
4 Town Farm Close
Oaklea, Honiton
Devon EX14 8YA
Tel: 01404 46872
Fax: 01404 44123
Contact: Kevin Holman
Number of Events: 10
Selection policy: A/B
Quality craft marquees at major county
shows representing a wide selection of
handmade British craftwork

QUALITY CRAFT & GIFT FAIRS
55 Alwoodley Lane
Alwoodley
Leeds
West Yorkshire LS17 7PU
Tel: 0113 267 1896
Fax: 0113 230 0225
Contact: Diane or Stuart Morris
Number of Events: 28-30
Selection policy: B
Crafts in hotels etc. in Yorkshire,
Durham, Cleveland and Nottinghamshire

QUINTET PROMOTIONS
4 Claremount Court
Dipe Lane
Boldon
Tyne & Wear NE36 0NF
Tel: 0191 536 2684
Contact: Margery Walsh
Number of Events: 30
Selection Policy: B
Quintet are organisers of quality fairs in
Northern England and Southern Scotland

REDBRIDGE SHOW
Show Office, PQD Division
5th Floor Front, Lynton House
255-259 High Road
Ilford, Essex IG1 1NY
Tel: 0181-478 9184
Fax: 0181-478 9129
Contact: Caroline Bernard
Number of Events: 1
Selection policy: B
Major show held in Ilford usually attracts
around 50,000 visitors

ROMOR EXHIBITIONS LTD
PO Box 448
Bedford
Bedfordshire MK40 2ZP
Tel: 01234 345715
Fax: 01234 328604
Contact: Anthony Rose
Number of Events: 28
Selection policy: B
Organising for 20 years in Bedfordshire,
Cambridgeshire, Northamptonshire,
Leicestershire and Buckinghamshire

ROYAL INTERNATIONAL AIR TATTOO
Building 15
RAF Fairford
Gloucestershire GL7 4DL
Tel: 01285 713300
Fax: 01285 713268
Contact: Tom Watts
Number of Events: 1
Selection policy: B
Giant event held at RAF Fairford attracts
around 150,000 visitors

RURAL CRAFTS ASSOCIATION
Brook Road
Wormley
Godalming
Surrey GU8 5UA
Tel: 01428 682292
Fax: 01428 685969

Contact: Trevor Sears or Karen Hall-
Sears
Number of Events: 56
Selection policy: A/B
Craft marquees at most major shows and
established Crafts for Christmas

SHACKLETONS
49 St Michaels Way
Partridge Green
Horsham
West Sussex
RH13 8LA
Tel/Fax: 01403 711066
Number of events: about 6
Selection Policy: B
These are professionally run shows with
the personal touch and a sense of
humour

SKILLBANK RESOURCES LTD
PO Box 35
Newtown
Powys SY16 4ZZ
Tel: 01686 670222
Fax: 01686 670222
Contact: Malcolm MacIntyre-Read
Number of Events: 3
Selection policy: B
Takes quality crafts to the NEC and out to
France

**SOUTHERN & SUSSEX CRAFT
EXHIBITIONS**
1 Falmer Court
London Road
Uckfield
East Sussex
TN22 1HN
Tel: 01825 766110
Fax: 01825 766112
Contact: Jacqui Giles
Number of Events: 4
Selection policy: B
Many exhibitors on show at fairs in
Brighton and Syon Park, Brentford

THE CRAFT MOVEMENT
PO Box 1641
Frome
Somerset BA11 1YY
Tel: 01373 813333
Contact: Victoria or Paula Blight
Number of Events: 12
Selection policy: A
Quality craftworkers at well regarded
series of events around England

THE CRAFT PEOPLE
Front Street
Ulceby
North Lincolnshire DN39 6SY
Tel: 01469 588774/0585 226042
Contact: Les Hutchinson
Number of Events: 24
Selection policy: C
Craft marquees from Kent/London right
up to Hull and York

THE EXHIBITION TEAM LTD
Events House
Wycombe Air Park
Marlow
Buckinghamshire SL7 3DP
Tel: 01494 450504
Fax: 01494 450245
Contact: Sue Lloyd
Number of Events: 30
Selection policy: B
Live crafts all around England

THE MALCOLM GROUP EVENTS LTD
Ground Floor
3 Brunswick Place
Hove
East Sussex BN3 1EA
Tel: 01273 723249
Fax: 01273 723249
Contact: Clive Geisler
Number of Events: 1
Selection policy: C
Crafts at the Medieval Festival at
Herstmonceux Castle in Sussex

TOWN AND COUNTRY CRAFT FAIRS
Hill Cross
Ashford
Bakewell
Derbyshire DE45 1QL
Tel: 01629 812008
Fax: 01234 262702
Number of Events: 20
Selection policy: B
Contact: Pat Paulett
Organisers of quality craft and teddy
bear fairs at Buxton, Cheltenham and
Midland game fair

TRICIA LEIGH EXPOSITIONS
37 Parlaunt Road
Langley
Slough
Berkshire
SL3 8BD
Tel: 01753 545384

Fax: 01753 545384
Contact: Tricia Leigh
Number of Events: 7
Selection policy: C
Racecourse based shows at Epsom and
Kempton Park – Surrey and Middlesex

TRUMPTON CRAFT FAYRES
Branson House
The Street
Walsham-le-Willows
Suffolk IP31 3AZ
Tel: 01359 258369
Fax: 01359 258370
Contact: Lisa Squire
Number of Events: 10
Selection policy: B
Relatively new organiser launching major
fayres in Suffolk

WALES CRAFT COUNCIL LTD
Park Lane House
7 High Street
Welshpool
Powys SY21 7JP
Tel: 01938 555313
Fax: 01938 556237
Contact: Helen Davies or Janet Edwards
Number of Events: 4
Specialises in taking Welsh crafts out to a
wider public

WEST COUNTRY CRAFT FAIRS
27 Jocelyn Drive
Wells
Somerset BA5 2ER
Tel: 01749 677049/0860 710693
Fax: 01749 677049
Contact: Fred Wilcox
Number of Events: 26
Selection policy: B/C
Not just West Country, but also Oxford-
shire, Warwickshire and Hampshire

WOODLAND CRAFTS
1 St John's Avenue
Purbrook
Waterlooville
Hampshire PO7 5PJ
Tel: 01705 614177
Fax: 01705 614177
Contact: Paul Bishop
Number of Events: 25-30
Selection policy: B
Concentrating on the Gosport and
Portsmouth area; also New Forest,
Lancing and Crawley

Abercraf Guild of Weavers, Spinners & Dyers
(Val Thomas)
24 Tan-y-Garth
Abercraf, Swansea
West Glam SA9 1XX

Aberdare Spinners
(Mrs Joan Le Carpentier)
25 + 30 The Market Hall
Aberdare
Mid Glam CF44 7EB
Tel: 01685 876759

Abergavenny Craft Cooperative
(Margaret & John Harvey)
Ingleby Cottage
Croseonen Road
Abergavenny
Gwent NP7 6AD
Tel: 01873 859506

Abergele Guild of Weavers, Spinners & Dyers
(Hilary Castle)
Bryn Eglwys, Pontygwyddel
Llanfair Talhaiarn
Clwyd
Tel: 01745 540360

An Bord Tráchtála
10 Claremont Terrace
Glasgow
Strathclyde G3 7XR
Tel: 0141-332 3015

Anglesey Craftworkers Guild
(A Person)
Glan Alaw
Llanddeusant
Holyhead, Anglesey
Gwynedd LL65 4AG
Tel: 01407 730601

Anglesey Embroiderers
(Mrs Margaret Hughes)
1 Bryntirion
Llanfairpwll, Anglesey
Gwynedd LL61 5YP
Tel: 01248 712084

Art & Design Movement
(Mrs J Darnley)
54 The Piece Hall
Halifax
West Yorks HX1 1RE

Art Workers Guild
(Hugh Krall)
6 Queens Square
London WC1N 3AR
Tel: 0171-837 3474

Arts Council of England
14 Great Peter Street
London SW1P 3NQ
Tel: 0171-333 0100

Arts Council of Wales Craft Department
(Roger Lefevre)
9 Museum Place
Cardiff
South Glam CF1 3NX
Tel: 01222 394711

Association Intercraft
(Colin Kentish)
33 rue du 4 Septembre
31220 Cazeres-sur-Garonne
FRANCE
Tel: +33 561-96-41-22

Association for Applied Arts
(Mandy Lee)
6 Darnaway Street
Edinburgh
Midlothian EH3 6BG
Tel: 0131-220 5070

Association of Rag Rug Makers
(Ann Davies)
1 Wingrad House
Jubilee Street
London
E1 3BJ
Tel: 0171-790 1093

Association of Artists in Ireland
Room 803
Liberty Hall, Dublin 1
EIRE
Tel: +353 1-874-0529

Association of Guilds of Weavers, Spinners & Dyers
(Anne Dixon)
2 Bower Mount Road
Maidstone
Kent
ME16 8AU
Tel: 01622 678429

Association of Illustrators
First Floor
32-38 Saffron Hill
London EC1N 8FH
Tel: 0171-831 7377

Association of Photographers
9-10 Domingo Street
London EC1 0TA
Tel: 0171-608 1441

Association of Visual Artists in Wales
PO Box 283, Cardiff
South Glam CF2 2YT

Association of Woodturners of Great Britain
(Hugh O'Neill)
Myttons Cottage, Boraston
Tenbury Wells
Worcs WR15 8LH
Tel: 01584 810266

Basketmakers' Association
(Mrs Ann Brooks)
Pond Cottage, North Road
Chesham Bois, Amersham
Bucks HP6 5NA
Tel: 01494 726189

Bath & District Craft Circle
(Mrs G Twort)
47 Bradford Park
Combe Down
Avon BA2 5PR
Tel: 01225 834966

Batik Guild
3 Vicarage Hill
Dadby, Daventry
Northants NN11 6AP

Bead Society of Great Britain
(Mrs Pat Nieburg)
Ventura, Horseshoe Road
Spalding, Lincs PE11 3BE
Tel: 01775 713330

Birmingham Calligraphy Society
(Nicholas Caulkin)
205 Dyas Avenue,
Great Barr, Birmingham
West Midlands B42 1HN
Tel: 0121-358 0032

**Birmingham Guild of
Weavers, Spinners & Dyers**
(Mrs Mence)
818 Warwick Road
Solihull
West Midlands B91 3HA
Tel: 0121-705 7236

Border Lacemakers
(Anne Stratton)
The Old Post Office
Penallt, Monmouth
Gwent NP5 4AH

Borders Craft Association
(Margaret Jeary)
Kalemouth, Kelso
Roxburghshire TD5 8LE
Tel: 01835 850266

**Brecknock Guild of
Spinners, Weavers & Dyers**
(Mrs Kelleher)
Upper Middle Road
Boughrood, Brecon
Powys LD3 0BX
Tel: 01874 754407

**Brecon & Radnor Guild of
Weavers, Spinners & Dyers**
(Patsy Beswick)
Guidfa House, Crossgates
Llandrindod Wells
Powys LD1 6RF
Tel: 01597 851241

Bridgend Quilters
(Mrs Norma Hutin)
Tel: 01656 653206

**British American Arts
Association**
(J Williams)
3rd Floor
116 Commercial Street
London E1 6NF
Tel: 0171-247 5385

**British Artist Blacksmiths
Association**
(Chris Topp), Lyndhurst
Carlton Husthwaite, Thirsk
North Yorks YO7 7BJ
Tel: 01845 501415

British Artists in Glass
Broadfield Glass Museum
Barnet Lane, Kingswinford
West Midlands DY6 9QA
Tel: 01384 273011

**British Ceramic
Confederation**
Federation House
Station Rd
Stoke-on-Trent
Staffs ST4 2SA
Tel: 01782 744631

**British China & Porcelain
Artists Association**
(Mrs E Bliss)
'High Dene'
Bramble Lane
Clanfield
Hants PO8 0RT

**British Chinese
Artists' Association**
Interchange Studios
Dalby Street, Kentish Town
London NW5 3NQ
Tel: 0171-267 6133

**British Decoy Wildfowl
Carvers Association**
(Alan Emmett)
6 Pendred Road
Reading
Berks RG2 8QL

**British Doll Artists
Association**
(June Rose Gale)
49 Cromwell Road
Beckenham
Kent BR3 4LL
Tel: 0181-658 1865

**British Jewellery &
Giftware Federation**
10 Vyse Street
Hockley
Birmingham
West Midlands B18 6LT
Tel: 0121-236 2657

British Stickmakers Guild
(Brian Aries)
44a Eccles Road
Chapel-en-le-Frith
High Peak
Derbys
SK12 6RG
Tel: 01298 815291

British Toymakers Guild
124 Walcot Street
Bath
Avon BA1 5BG
Tel: 01225 442440

**British Woodcarvers
Association**
(John Sullivan)
25 Summerfield Drive
Nottage, Porthcawl
Mid Glam CF36 3PB
Tel: 01656 786937

**British Woodcarvers
Association (Welsh Region)**
(Mrs S Littley)
32 Hafod Las
Pencoed
Mid Glam CF35 5NB

**Bucks Guild of Weavers,
Spinners & Dyers**
(Pam Mitchell)
Braziers Well, Braziers End
Chesham
Bucks HP5 2UL
Tel: 01240 29527

**Bucks Pottery and
Sculpture Society**
(Monica Plant)
8 Harvey Orchard
Beaconsfield
Bucks HP9 1TH

**Calligraphy & Lettering
Arts Society**
(Mrs Sue Cavendish)
54 Boileau Road
London SW13 9BL
Tel: 0181-741 7886

Cardiff Quilters
(Mrs Tricia Davies)
41 Cherry Orchard Road
Lisvane, Cardiff
South Glam CF4 5UE

**Ceredigion Guild of
Weavers,
Spinners & Dyers**
(Pat Nathan)
Maesteg
Rhydlewis
Llandysul
Dyfed SA44 5PT

**Clare Association of
Artists and Craftworkers**
(Deirdre O'Brien)
Wildflower Studio
Boston Tubber
County Clare
EIRE
Tel: +353 (0)91-633284

Cleddau Lacemakers
(Vera Robinson)
8 Castle Pill Crescent
Steynton, Milford Haven
Dyfed SA73 1HD

**Clwyd Guild of Weavers,
Spinners & Dyers**
(Mrs Judith Kirkham)
Bluebell House
Baguit, Clwyd CH6 6HE

**Contemporary Basketry
Group**
(Kate Reading)
Tan-Ffynon, Penuwch
Tregaron
Dyfed SY25 6RF
Tel: 01974 423434

**Contemporary Crafts
Network**
Pearoom Centre for
Contemporary Craft
Station Yard
Heckington, Sleaford
Lincs NG34 9JJ
Tel: 01529 460765

Cotswold Craftsmen
(Tamsin Marsh)
Buttermilk Farm
Toddington, Cheltenham
Glos GL54 5DG
Tel: 01242 621245

Country Crafts Association
(Colin Hornsey)
29 Wallis Avenue
Lincoln, Lincs LN6 8AS
Tel: 01522 687911

**Coventry & District Guild of
Weavers, Spinners & Dyers**
(Mrs H Eden)
14 St Osburg's Road
Coventry, Warks CV2 4EG
Tel: 01203 449999

Cowbridge Lacemakers
(Mrs B James)
196 Malin Crescent
Bridgend, Mid Glam

Craft Connection
(Margaret Gault)
62 Cushendall Road
Ballycastle
Co Antrim BT54 6QR
Tel: 01265 762130

**Craft Guild of
West Lancashire**
(Theresa Gaskell)
1 Westgate, Pennylands
Skelmersdale
Lancs WN8 8LP
Tel: 01695 50200

Craft Potters Association
(Marta Donaghey)
William Blake House
7 Marshall Street
London W1V 1LP
Tel: 0171-437 7605

Crafts Council
44a Pentonville Road
Islington
London N1 9BY
Tel: 0171-278 7700

Crafts Council of Ireland
Townhouse Centre
South William Street
Dublin 2 EIRE
Tel: 00-353-1-679-7383

Craftworks (NI) Ltd
(Mrs Patricia Flanagan)
17-19 Linenhall Street
Belfast BT2 8AA
Tel: 01232 236334

**Craven Guild of Weavers,
Spinners & Dyers**
(Mrs Beryl Cooke)
2 Woodlands Grove
Ilkley
West Yorks LS29 9BX
Tel: 01943 608483

**Crefftau'r Cymoedd
(Crafts of the Valleys)**
(Mrs S Kearns)
40 St Patrick's Drive
Bridgend
Mid Glam CF31 1RP

**Dacorum & Chiltern
Potters' Guild**
(Ruth Karnac)
35 Kingsend
Ruislip
Middx HA4 7DD
Tel: 01895 631738

Design Council
28 Haymarket
London
SW1Y 4SU

Design Workers Guild
(Melanie Adkins)
The Porch House
Swan Hill, Shrewsbury
Salop SY1 1NQ
Tel: 01743 241031

Designer Bookbinders
6 Queen Square
London WC1N 3AR
Tel: 01248 602591

Designer Jewellers Group
24 Rivington Street
London EC2A 3DU
Tel: 0171-739 3663

**Development Board for
for Rural Wales**
Ladywell House
Newtown
Powys
Tel: 01686 626965

Devon Guild of Craftsmen
(Francis Byng)
Riverside Mill
Bovey Tracey
Devon TQ13 9AF
Tel: 01626 832223

**Diss and District Guild of
Weavers, Spinners & Dyers**
(Mrs Pam Ross)
'Southview', Common Road
Shelfanger, Diss
Norfolk IP22 2DP
Tel: 01379 643563

Dorset Craft Guild
(Philip Goulden)
Walford Mill Craft Centre
Stone Lane, Wimborne
Dorset BH21 1NL
Tel: 01202 841400

Dorset Pottery Group
(Alan Ashpool)
Trumps In Cottage
Whitchurch Canonicorum
Bridport
Dorset DT6 6RH
Tel: 01297 489347

Dove Lacemakers
(Mrs Diana Millner)
44 Ferrers Avenue
Tutbury, Burton-on-Trent
Staffs DE13 9JR
Tel: 01283 812477

**Dry Stone Walling
Association of Gt Britain**
(Jacqui Simkins)
National Agricultural Centre
YFC Centre, Stoneleigh Park
Warks CV8 2LG
Tel: 0121-378 0493

**Dumfries & Galloway Guild
of Weavers, Spinners &
Dyers**
(Elizabeth Fellowes)
West Isle, Islesteps
Dumfries DG2 8ES
Tel: 01387 62094

**Dyfed Craft Community
Cooperative**
(David Petersen)
Derw Mill, Pentre Cwrt
Llandysul
Dyfed SA44 5DB

**East Anglian Potters
Association**
(Cathy Border)
37 Great Farthing Close
St Ives, Cambs PE17 4JX
Tel: 01480 466177

**East Devon Small
Industries Group**
(G P Hulley)
115 Border Road,
Heathpark, Honiton
Devon EX14 8BT
Tel: 01404 41806

East Midlands Arts Board
(Visual Arts Assistant)
Mountfields House
Epinal Way, Loughborough
Leics LE11 0QE
Tel: 01509 218292

**East Sussex Guild
of Craftworkers**
(Alf Case)
Little Clays, Willingford La
Burwash Weald, Etchingham
East Sussex TN19 7HR
Tel: 01435 882707

**East Sussex Guild of
Weavers, Spinners & Dyers**
(Mrs Sheelagh Hoblyn)
Great Streele Cottage
Framfield, nr Uckfield
East Sussex TN22 5SA
Tel: 01825 890425

**Edinburgh Guild of
Weavers,
Spinners & Dyers**
(Alison Midson)
85 Backmarch Road
Rosyth
Fife KY11 2RP
Tel: 01383 414514

**Edinburgh Knitting
& Crochet Guild**
(Isabella Ricketts)
3 Gillsland Road
Edinburgh
Midlothian
EH10 5BW
Tel: 0131-337 3984

**Egg Crafters Guild
of Great Britain**
(Joan Cutts)
The Studio
7 Hylton Terrace
North Shields
Tyne & Wear
NE29 0EE
Tel: 0191-258 3648

Embroiderers' Guild
Apartment 41
Hampton Court Palace
East Molesey
Surrey KT8 9AU
Tel: 0181-943 1229

**Embroiderers' Guild
Carmarthen**
(Mrs Esme Wagstaff)
Brodawel
Crwbin
Kidwelly
Dyfed SA17 5DE
Tel: 01269 870196

**Embroiderers' Guild
Mid Wales**
(June Morgan)
Pen-y-Bont
Gellilydan
Maentwrog
Gwynedd

**Embroiderers' Guild
North Wales**
(Mair Griffiths)
Pen y Bryn
Waen Wen
Bangor
Gwynedd
LL57 4UF

**Embroiderers' Guild
Pembrokeshire**
(Mrs Jo Jones)
5 Nestor Square
Narberth
Dyfed SA67 7UG
Tel: 01834 861528

**Embroiderers' Guild
South East Wales**
(Mrs Jones)
26 Waterloo Gardens
Penylan, Cardiff
South Glam CF2 5AB

**Embroiderers' Guild
Kingston Branch**
(Marina Roussac-Hatton)
156 Walsingham Gardens
Ewell
Surrey KT19 0NF
Tel: 0181-393 1864

**Embroiderers' Guild
Central West Study Group**
(Pamela Lee)
40 Bishops Park
Pembroke
Dyfed SA71 5JP

Embroiderers' Guild Gwent
(Alicia Badby)
Ty-Gwyn House
Llantilio Pertholey
Abergavenny
Gwent NP7 6NY
Tel: 01873 856356

**Embroiderers' Guild
Swansea**
(Mrs Beverley Boucher)
99 Kingrosia Park
Clydach
Swansea
West Glam SA6 5PJ
Tel: 01792 842738

Essex Craft Society
(Lesley Williams)
52 Sixth Street
Chelmsford
Essex
CM1 4ED

Farnborough Craft Guild
(Mrs Jan Strode)
9 Hart Close
Farnborough
Hants GU14 9HQ
Tel: 01276 31926

**Federation of
British Artists**
(John Sayers)
17 Carlton House Terrace
London
SW1Y 5BD
Tel: 0171-930 6844

Fife Craft Association
(Peter Leigh)
12 Valley Grove
Leslie, Glenrothes
Fife KY6 3BZ
Tel: 01592 743539

**Flower and
Plants Association**
New Covent Garden Market
London SW8 5NX

**Glamorgan Guild of
Weavers, Spinners & Dyers**
(Jean Roberts)
4 Heol Don
Whitchurch
Cardiff
South Glam
Tel: 01222 627326

Glamorgan Quilters
(Val Spierling)
Swn y Nant
75 Nant Talwg Way
Barry
South Glam CF6 8LZ

**Gloucestershire Guild
of Craftsmen**
(Tony Davies)
Bredon Pottery, High Street
Bredon
Tewkesbury
Glos GL20 7LW
Tel: 01684 773417

Goldsmith's Craft Council
(Malcolm Pullan)
Goldsmith's Hall
Foster Lane
London
EC2V 6BN
Tel: 0171-606 7010

**Grafton Plyford
Craft Group**
(Sue Johnson)
11 Wirehill Drive
Lodge Park
Redditch
Worcs B98 7JU

**Grampian Guild of
Spinners, Weavers & Dyers**
(Carole Keepax)
Auchravie, Monymusk
Inverurie
Aberdeenshire AB51 7SQ
Tel: 01467 651314

Guild of Artisans
(Colin Crawley)
The Cottage, West End Road
West End, Southampton
Hants SO30 3BH
Tel: 01703 477711

Guild of British Découpeurs
(Mary Lewis)
The Cottage
Barton End House, Bath Road
Nailsworth, Glos GL6 0QQ
Tel: 01453 833465

**Guild of Disabled
Homeworkers**
Enterprise Aid Centre
Stag House
Woodchester
Glos GL5 5EZ
Tel: 01453 835623

Guild of Enamellers
(Kathleen Kay)
8 Himley Avenue
Dudley
West Midlands DY1 2QP
Tel: 01384 256438

Guild of Glass Engravers
(Mrs Gail Plant)
19 Wildwood Road
London
NW11 6UL
Tel: 0181-731 9352

**Guild of Herefordshire
Craftsmen**
(Emma Baker)
Castle Weir
Lyonshall, Kington
Herefordshire
HR5 3HR
Tel: 01544 340332

**Guild of Lincolnshire
Craftsmen**
(D Duncombe)
Kingshill Cottage
10 Cheviot Street
Lincoln
Lincs LN2 5JD

Guild of Master Craftsmen
(John Kilroy)
166 High Street
Lewes
East Sussex BN7 1XU
Tel: 01273 477374

Guild of Needle Lacers
72 Hawes Road
Northwood
Middx HA6 1EW
Tel: 01927 429844

Guild of Sussex Craftsmen
(Judith Fisher)
Huntswood
St Helena's Lane
Streat, Hassocks
West Sussex
BN6 8SD
Tel: 01273 890088

**Guild of West Midlands
Artists & Craftsmen**
(Christine Parkes)
9 Squires Croft
Sutton Coldfield
West Midlands
B76 8RY
Tel: 0121-351 4330

**Guild of Wrought
Ironsmiths in Wales**
(Emlyn Thomas)
Economic Development Unit
County Hall, Aberaeron
Dyfed

**Gwent Guild of Weavers,
Spinners & Dyers**
(Mrs Sheila Morgan)
20 St Mary's Close
Griffithstown
Pontypool
Gwent NP4 5LS
Tel: 01495 753255

Gwent Quilters
(Maureen de Souza)
27 Old Market Street
Usk, Gwent

**Gwynedd & Clwyd
Association of Craftworkers**
(Joy Harper)
Awel Mor, Mostyn Road
Greenfield
Clwyd
CH8 9DN
Tel: 01352 711539

**Gwynedd Guild of
Weavers, Spinners & Dyers**
(Jan Rowlands)
Plas Iolyn, Llangoed
Anglesey, Gwynedd

**Heart of England
Craft Workers' Guild**
(Chris/Bettie Carless)
105 St Georges Lane
Worcester
Worcs WR1 1QS
Tel: 01905 29285

**Herefordshire Guild of
Weavers, Spinners & Dyers**
(Judith Baresel)
56 Broomy Hill
Hereford
Herefordshire HR4 0LQ
Tel: 01432 269073

**International Feltmakers
Association**
(Ewa Kuniczak)
23 Glebe Road
Kincardine-on-Forth, Alloa
Clackmannanshire
FK10 4QB
Tel: 01259 730779

**Isle of Wight Guild of
Weavers, Spinners & Dyers**
(R Philips)
2 Harbour Villas
Newlands, St Helens
Isle of Wight
Tel: 01983 873960

Ivanhoe Crafts Guild
(J M Butler)
15 Roxby Close, Doncaster
South Yorks DN4 7JH
Tel: 01302 539104

**Kennet Valley Guild of
Weavers, Spinners & Dyers**
(A J Moss)
71 Maple Crescent
Newbury
Berks RG13 1LP
Tel: 01635 43846

**Kent Guild of Spinners,
Dyers & Weavers**
(Mrs J Wood)
Lords Spring Farm Cottage
Bitchet Green, nr Sevenoaks
Kent TN15 0NA
Tel: 01732 762963

Kent Potters Association
(Janet Jackson)
Fairview Barn, Upper Street
Broomfield, nr Maidstone
Kent ME17 1PS
Tel: 01622 863554

Kingston Weavers
Church Farm
Kingston St Mary
Taunton, Somerset
Tel: 01823 45267

Knitting & Crochet Guild
(Mrs Anne Budworth)
228 Chester Road North
Kidderminster
Worcs DY10 1TH
Tel: 01562 754367

Knitting Craft Group
(Alec Dalglish)
Market Place
Educational Service
37 Market Place
Thirsk
North Yorks
YO7 1HA
Tel: 01845 524300/537280

Lace Guild
(Miss Gwynedd Roberts)
'The Hollies'
53 Audnam
Stourbridge
West Midlands DY8 4AE
Tel: 01384 390739

Ladybug Quilters
(Mrs O Pavett)
Summer Leas
13 Gloucester Close
Llanyrafon
Cwmbran
Gwent
Tel: 01633 869049

Llantwit Major Lace Circle
18 Cardigan Crescent
Boverton
Llantwit Major
South Glam

**Llyn Guild of Weavers,
Spinners & Dyers**
(Eirian Larsen)
48 Ffordd Eryn
Caernarfon
Gwynedd LL55 2UR
Tel: 01286 672353

**London & Home Counties
Guild of Weavers Spinners
& Dyers**
(Miss Elizabeth Jackson)
13 Coverdale Road
London W12 8JJ
Tel: 0181-740 6832 (eve)

London Potters
(Mary Lambert)
105 Albert Bridge Road
London SW11
Tel: 0171-228 7831

Made in Scotland Ltd
(Peter Guthrie)
The Craft Centre
Station Road
Beauly
Inverness-shire IV4 7EH
Tel: 01463 782578

Makers Guild in Wales
(Lauraine Voisey)
Craft in the Bay
72 Bute Street, Cardiff Bay
South Glam
CF1 6AX
Tel: 01222 491136/484611

**Marquetry Society
West Wales Group**
(Mrs P M Austin)
The Barn House
Llanon
nr Aberystwyth
Dyfed SY23 5LZ
Tel: 01974 202581

**Meirionnydd Needlecrafts
Group**
(Christine Bursnall)
Tan y Buarth
Old Llanfair Road, Llanfair
Gwynedd LL46 2SY

**Midland Handweavers'
Association**
(P H M Butcher)
61 Onibury Road
Handsworth
Birmingham
West Midlands B21 8BE

Midland Potters Association
(Keith Cherry)
104 Shakespeare Drive
Nuneaton
Warks CV11 6NW
Tel: 01203 329114

Milton Keynes Craft Guild
(Clare Layton)
Saxon Court
502 Avebury Boulevard
Central Milton Keynes
Bucks MK9 3HS
Tel: 01908 694764

Milton Keynes Printmakers
(Myriam Metcalf)
South Pavilion
Gt Linford Art Centre
Parklands
Great Linford
Bucks MK17 8SP
Tel: 01908 583845

Montgomery Guild of Weavers, Spinners & Dyers
(Mrs S Race)
Lluest Carno
Caersws
Powys SY17 5LX

National Acrylic Painters' Association
(Kenneth Hodgson)
134 Rake Lane
Wallasey, Wirral
Merseyside L45 1JW
Tel: 0151-639 2980

National Artists Association
Spitalfields
21 Steward Street
London E1 6AJ
Tel: 0171-426 0911

National Association of Disabled Craft Workers
(Alex Conrade-Marshall)
Piethorn Cottage
Barrachan by Mochrum
Newton Stewart
Wigtownshire DG8 9NF
Tel: 01988 860204

National Association of Goldsmiths
78a Luke Street
London EC2A 4PY
Tel: 0171-613 4445

National Clayware Federation
Aberdeen House
Heronsgate Road
Chorleywood
Herts WD3 5BB

National Patchwork Association
PO Box 300
Hethersett, Norwich
Norfolk NR9 3DB
Tel: 01603 812259

New Forest Crafts Association
The Lodge
Mill Lane, Burley
Hants BH24 4HP

Norfolk & Suffolk Guild of Weavers, Spinners & Dyers
(Carolyn Reeder)
Warren Anne
Flixton
Suffolk

Norfolk Contemporary Craft Society
(Mrs B Prior)
Gilderswood Farm
Forncett St Peter, Norwich
Norfolk NR16 1LN
Tel: 01953 789362

Norfolk Craftsmen's Guild
(Mrs Rosemary Kingsland)
Rosecroft
Rode Lane, Carleton Rode
Norwich
Norfolk
NR16 1NW
Tel: 01953 860706

North Wales Potters
(Steve Mattison)
Cae Carrog
Aberhosan, Machynlleth
Powys SY20 8SE
Tel: 01654 703247

North West Arts Board Craft Marketing Initiative
Drumcroon Arts Centre
2 Parsons Walk
Wigan
Lancs WN1 1RS
Tel: 01942 825088

Northamptonshire Guild of Designer Craftsmen
(Bob Walder)
28 High Street
Milton Malsor
Northants NN7 3AS
Tel: 01604 858470

Northern Potters Association
(Chris Utley)
West Wood House
Sutton-on-Derwent
North Yorks YO4 5BT
Tel: 01904 608409

Nottinghamshire County Crafts
(Mrs J Fairgrieve)
11 Eton Grove
Wollaton Park
Notts NG8 1FT
Tel: 0115 928 3431

Notts & District Guild of Spinners, Weavers & Dyers
(Glennis Eaton)
47 Church Street
Lambley
Notts NG4 4QB
Tel: 0115 931 2804

Ochil Craft Association
(L Richmond)
28 Grant Street, Alloa
Clackmannanshire
FK10 1ND
Tel: 01259 213403

Origin Dyfed Craft Cooperative
1 St Mary's Street
Camarthen, Dyfed SA31 1JT
Tel: 01267 238507

Origin Gwynedd
(Melina Eryri Cyf)
Snowdon Mill
Snowdon Street
Porthmadog
Gwynedd LL49 9DF
Tel: 01706 512137

Outer Hebrides Craft Association
(Susan Robson)
2 Holm, nr Stornoway
Isle of Lewis
Western Isles PA86 0AZ

Oxford Guild of Weavers, Spinners & Dyers
(Gladys May)
21 Abingdon Road
Cumnor
Oxon OX2 9QN
Tel: 01865 864050

Oxfordshire Craft Guild
(Valerie Newey)
c/o The Craft Shop
7 Goddards Lane
Chipping Norton
Oxon OX7 5NP
Tel: 01608 641525

**Paperweight
(Group for Paper
Makers & Paper Artists)**
(Polly Blythin)
99 Colebrook Street
Winchester
Hants SO23 9LH
Tel: 01962 865993

**Pembrokeshire Craftsman's
Circle**
(Hilary Bassett)
Bryn Pottery
Eglwyswrw
Crymych
Dyfed SA41 3SS
Tel: 01239 891608

**Pembrokeshire Guild of
Weavers, Spinners & Dyers**
(Mrs Sue Hoyland)
Rose Cottage
Sandy Hill Rd
Saundersfoot
Dyfed SA69 9HN
Tel: 01834 812544

Perfumers Guild Ltd
(John Bailey)
61 Abbots Road
Abbots Langley
Herts WD5 0BJ
Tel: 01923 260502

Poole Printmakers
5 Bowling Green Alley
Poole
Dorset BH15 1AG
Tel: 01202 393776/734521

Pressed Flower Craft Guild
(Joyce Fenton)
41 The Street
Charlwood, Horley
Surrey

Printmakers Council
(Kieron Farrow)
Clerkenwell Workshops
31 Clerkenwell Close
London EC1R 0AT
Tel: 0171-250 1927

Quilt Art
(Sue Hagley)
Garage Cottage
Norwich Road
Barham
nr Ipswich
Suffolk IP6 0PP
Tel: 01473 832503

Quilters' Guild
(The Administrator)
Room 190
Dean Clough
Halifax
West Yorks
HX3 5AX
Tel: 01422 347669/345017

**Rhymney Valley
Lacemakers**
(Mrs M Gardner)
Y Blanfa
Nelson Road
Ystrad Mynach
Mid Glam CF8 7EG

Royal Photographic Society
Milson Street
Bath
Avon BA1 1DN
Tel: 01225 462841

Royal Society
6 Carlton House Terrace
London
SW1Y 5AG

Royal Watercolour Society
(Eleanor Blake)
Bankside Gallery
48 Hopton Street
Blackfriars
London SE1 9JH
Tel: 0171-928 7521

Rugby Craft Association
(Mrs C E Hughes)
10 Dewar Grove
Hillmorton
Rugby
Warks CV21 4AT
Tel: 01788 575761

**Rural Development
Commission**
141 Castle Street
Salisbury
Wilts
SP1 3TP
Tel: 01722 336255

Scottish Arts Council
(Monica Roscrow)
Crafts Department
12 Manor Place
Edinburgh
Midlothian EH3 7DD
Tel: 0131-243 2437

**Scottish Potters
Association**
(Maggie Longstaff)
'Murrayfield'
Roslin Glen, Roslin
Midlothian
EH25 9PY
Tel: 0131-440 2228

**Sculptors' Society
of Ireland**
(Aisling Prior)
119 Capel Street
Dublin 1, EIRE
Tel: (+353) 1-8722296

Sheffield Craft Guild
Limestone Hall Farm
Limestone Cottage Lane
Wadsley Bridge
Sheffield
South Yorks S6 1NJ
Tel: 0114 285 3775

Sirhowy Valley Lacemakers
(Mrs B Bridge)
Anakana, Le Julian Heights
Fleur-de-Lys
Blackwood
Gwent

**Skye & Lochalsh Arts
and Crafts Association**
(Stuart Whalley)
Edinbane Pottery
Edinbane
Isle of Skye IV51 9PW
Tel: 01470 582234

**Society for Italic
Handwriting**
(Nicholas Caulkin)
205 Dyas Avenue
Great Barr, Birmingham
West Midlands B42 1HN
Tel: 0121-358 0032

Society of Amateur Artists
(John Hope-Hawkins)
PO Box 50, Newark
Notts NG23 5GY
Tel: 01949 844050

Society of Architect Artists
(Jo Burnham)
Capital House
25 Chapel Street
London NW1
Tel: 0171-262 3484

Society of Craftsmen, Kemble Gallery
(Mrs C Bulmer)
29 Church Street
Hereford
Herefordshire HR1 2LR
Tel: 01432 266049

Society of Designer Craftsmen
(Honorary Secretary)
24 Rivington Street
London EC2A 3DU
Tel: 0171-739 3663

Society of Graphic Fine Art
15 Willow Way
Hatfield
Herts AL10 9QD

Society of Northumbrian Craftsmen
(Barbara Scoins)
53/55 Bath Terrace
Newcastle upon Tyne
Tyne & Wear NE3 1UJ
Tel: 0191-285 8558

Society of Portrait Sculptors
(David Houchin)
27 Winchester Street
London
W3 8PA
Tel: 0181-992 7279

Society of Scottish Artists
(Anne Wishart)
69 Promenade
Portobello
Edinburgh
EH15 2DX
Tel: 0131-669 0637

Somerset Guild of Craftsmen
(Ron Rudd)
92 Alfoxton Road
Bridgwater
Somerset
TA6 7NW
Tel: 01278 424983

South East Arts Board
(Anne Satow)
10 Mount Ephraim
Tunbridge Wells
Kent
TN4 8AS
Tel: 01892 515210 Ext 214

South Wales Lacemakers
(Mrs J Turner)
34 Hafod Las, Pencoed
Bridgend
Mid Glam
CF35 5NB

South Wales Potters
(Cheryl Kirby)
Newtown Cottage
Lower Eggleton
near Ledbury
Herefordshire
HR8 2TZ

South West Arts
(Crafts Administrator)
Bradninch Place
Gandy Street
Exeter
Devon EX4 3LS
Tel: 01392 218188

South West Branch of Guild of Glass Engravers
(Chris Ainslie)
105 Woodmancote
Yate, Bristol
BS17 4LH
Tel: 01454 316110

South West Textile Group
(Teresa Searle)
48 Monmouth Road
Bishopstown, Bristol
BS7 8LG
Tel: 0117 924 7836

Southern Arts
(Crafts Officer)
13 St Clement Street
Winchester
Hants SO23 9DQ
Tel: 01962 855099

Southern Ceramic Group
(Martha Bowles)
24 Stubbington Way
Fair Oak
Eastleigh
Hants
Tel: 01703 629970

St Mary's Guild of Weavers, Spinners & Dyers
(Mrs S Fawcus)
2 Rose Villa, Anchor Road
Spa Common
Norfolk NR28 9AJ
Tel: 01692 405917

Strait Lacers
(Mrs S Cooke)
Llys Myfyr
Llansadwrn, Menai Bridge
Gwynedd LL59 5SL
Tel: 01248 811437

Suffolk Craft Society
(Monique Gregson)
Bridge Green Farm
Gissing Road
Burston, Diss
Norfolk IP22 3UD
Tel: 01379 740528

Surrey Guild of Craftsmen
1 Moushill Lane
Milford
Godalming
Surrey GU8 5BH
Tel: 01483 424769

Tawe Guild of Weavers & Spinners
(Cath Gallagher)
32 Ynysmeudwy Road
Ynysmeudwy, Pontardawe
West Glam SA8 4QD
Tel: 01792 865482

Textile Society
(Walter Bowyer)
173 Brettenham Road
Walthamstow
London E17 5AX
Tel: 0181-523 2399

The Craft Guild - Designer Makers in Bath & Bristol
(Jill Bartlett)
2 Bathwick Terrace
Bath, Avon BA2 4EL
Tel: 01225 461825

Ulster Guild of Weavers, Spinners & Dyers
(Pamela Dunbar)
123 Belsize Road
Lisburn
Co Antrim
BT27 4BT
Tel: 01846 674365

Vale of Glamorgan Batik Guild
(Ann Lewis)
3 Talyfan Close
Cowbridge
South Glam

Wales Craft Council
(Janet Edwards)
Park Lane House
7 High Street
Welshpool
Powys SY21 7JP
Tel: 01938 555313

Watercolour Society of Wales
(Margaret Butler)
4 Castle Road
Raglan
Gwent NP5 2JZ
Tel: 01291 690260

Welsh Development Agency
Llys Garth
Garth Road
Bangor
Gwynedd LL57 2RT

Wessex Guild of Craftsmen
(Allan Parsons)
91 Green Lane
Clanfield
Hants PO8 0LG
Tel: 01705 571744

West Essex Guild of Weavers, Spinners & Dyers
(Mrs Sue Prior)
Buckler Hall Farm
Perry Green
Much Hadham
Herts
SG10 6EA
Tel: 01279 842334

West Midlands Arts
Penny Smith –Crafts Officer
82 Granville Street
Birmingham
West Midlands B1 2LH
Tel: 0121-631 3121

West Wales Lacemakers
(Mrs K Joseph)
33 Mysydd Road
Landore, Swansea
West Glam
SA1 2NZ

Worcestershire Guild of Weavers, Spinners & Dyers
(Jean Burdett)
Hoe Court, Mathon Road
Colwall
Worcs
Tel: 01684 40321

World Crafts Council/Europe
(Katrin Strauss)
Rheinstrasse 23
60325 Frankfurt am Main
GERMANY
Tel: +49 (0)69-743-2113

Worshipful Company of Basketmakers
(Mary Butcher)
6 Downs Road
Canterbury
Kent
CT2 7AY
Tel: 01227 766427

Worshipful Company of Glaziers & Painters of Glass
(P Batchelor)
Glaziers Hall
9 Montague Close
London
SE1 9DD
Tel: 0171-403 3300

Wyre Craft Guild
(Edith Would)
Mount Pavilion
Fleetwood
Lancs

York & District Guild of Weavers, Spinners & Dyers
(Mrs Enid Parker)
Weavery
2 New Road
Brandesburton, Driffield
East Yorks
YO25 8RX
Tel: 01964 543123

Art & Craft Magazine
Scholastic Ltd
Villiers House
Clarendon Avenue
Leamington Spa
Warks
CV32 5PR

Artists Newsletter
AN Publications
PO Box 23
Sunderland
Tyne & Wear
SR4 6DG
Tel: 0191-514 3600
Fax: 0191-564 1600
Contact: Julie Crawshaw

Beautiful Stitches
Maze Media Ltd
Castle House
97 High Street
Colchester
Essex CO1 1TH
Tel: 01206 571385
Fax: 01206 571607

British Blacksmith
British Artist Blacksmiths
Association
(Chris Topp)Lyndhurst
Carlton Husthwaite
Thirsk
North Yorks YO7 7BJ
Tel: 01845 501415

Ceramic Review
21 Carnaby Street
London W1V 1PH
Tel: 0171-439 3377
Fax: 0171-287 9954

Ceramics Monthly
American Ceramic Society
PO Box 6136
Westerville
OH 43086-6136
USA

Craft Review
Crafts Council of Ireland
Townhouse Centre
South William Street
Dublin 2 EIRE
Tel: 00-353-1-679-7383

Crafts Beautiful
Maze Media Ltd
Castle House
97 High Street
Colchester
Essex
CO1 1TH
Tel: 01206 571385
Fax: 01206 571607

Crafts Magazine
Crafts Council
44a Pentonville Road
Islington
London
N1 9BY
Tel: 0171-278 7700
Fax: 0171-837 6891

Creative Ideas for the Home
GMC Publications Ltd
Castle Place
166 High Street
Lewes
East Sussex BN7 1XU
Tel: 01273 477374
Fax: 01273 478606

Crefft
Arts Council of Wales
9 Museum Place
Cardiff
South Glam
CF1 3NX
Tel: 01222 394711

Cross Stitcher
Future Publishing Ltd
Beauford Court
30 Monmouth Street
Bath
Avon
BA1 2BW
Tel: 01225 442244
Fax: 01225 462986

Design Business
Business Design Centre
52 Upper Street
Islington Green
London
N1 0QH
Tel: 0171-359 3535
Fax: 0171-226 0590

Doll Magazine
Ashdown Publishing Ltd
Avalon Court
Star Road
Partridge Green
West Sussex RH13 8RY
Tel: 01403 711511
Fax: 01403 711521

Dolls House World
Ashdown Publishing Ltd
Avalon Court
Star Road
Partridge Green
West Sussex
RH13 8RY
Tel: 01403 711511
Fax: 01403 711521

Engineering in Miniature
TEE Publishing
The Fosse
Fosse Way
Radford Semele
nr Leamington Spa
Warks
CV31 1XN
Tel: 01926 614101
Fax: 01926 614293

Furniture & Cabinetmaking
GMC Publications Ltd
Castle Place
166 High Street
Lewes
East Sussex BN7 1XU
Tel: 01273 477374
Fax: 01273 478606

Gifts International
Nexus Media
Nexus House
Swanley
Kent BR8 8HY

Gifts Today
Lema Publishing Co
Unit No 1
Queen Mary's Avenue
Watford
Herts WD1 7JR
Tel: 01923 250909
Fax: 01923 250995

Good Woodworking
Future Publishing Ltd
Beauford Court
30 Monmouth Street
Bath
Avon BA1 2BW
Tel: 01225 442244
Fax: 01225 462986

Housewares
Miller Freeman PLC
Sovereign Way
Tonbridge
Kent TN9 1RW
Tel: 01732 364422
Fax: 01732 361534

Inspirations
GE Publishing Ltd
133 Long Acre
London WC2E 9AW
Tel: 0171-836 0519
Fax: 0171-836 0280

International Dolls House
Nexus Special Interests
Nexus House
Boundary Way
Hemel Hempstead
Herts HP2 7ST
Tel: 01442 66551

**Journal for Weavers,
Spinners & Dyers**
Crockers
Rushey Way
Lower Earley
Berks RG6 4AS
Tel: 01734 867589

Knitting & Haberdashery
The Needlecrafts Review
Bates Business Centre
Church Road
Harold Wood
Romford
Essex
RM3 0JF
Tel: 01708 379897
Fax: 01708 379804

Knitting Machine Journal
Hawthornes
Whitecroft, nr Lydney
Glos
GL15 4PF
Tel: 01594 562161

Makers' News
Crafts Council
44a Pentonville Road
Islington
London
N1 9BY
Tel: 0171-278 7700
Fax: 0171-837 6891

Needlecraft
Future Publishing Ltd
Beauford Court
30 Monmouth Street
Bath
BA1 2BW
Tel: 01225 442244
Fax: 01225 462986

**Needlecrafts Cross Stitch
Collection**
Future Publishing Ltd
Beauford Court
30 Monmouth Street
Bath
BA1 2BW
Tel: 01225 442244
Fax: 01225 462986

New Stitches
Creative Crafts Publishing
The Old Grain Store
Brenley Lane
Boughton-under-Blean
Faversham, Kent
ME13 9LY
Tel: 01227 750215

**Passap Knitting Machine
Journal**
Hawthornes,
Whitecroft, nr Lydney   ·
Glos GL15 4PF
Tel: 01594 562161

Popular Crafts
Nexus Special Interests
Nexus House
Boundary Way
Hemel Hempstead
Herts HP2 7ST
Tel: 01442 66551

Perfect Home Magazine
Times House
Station Approach
Ruislip
Middx HA4 8NB

**Practical Craft Magazine
Magmaker Ltd
Cromwell Court
New Road, St Ives
Cambs PE17 4BG
Tel: 01480 496130
Fax: 01480 495514
Contact: Peter Raven**

Practical Woodworking
IPC Magazines
Kings Reach Tower
Stamford Street
London SE1 9LS
Tel: 0171-261 6689
Fax: 0171-261 7555

Progressive Gifts
Max Publishing Ltd
United House, North Road
London N7 9DP
Tel: 0171-700 6740
Fax: 0171-609 4222

Routing
Nexus Special Interests
Nexus House
Boundary Way
Hemel Hempstead
Herts HP2 7ST
Tel: 01442 66551

6000 Journal
Hawthornes
Whitecroft, nr Lydney
Glos GL15 4PF
Tel: 01594 562161

Slipknot
Knitting & Crochet Guild
228 Chester Road North
Kidderminster
Worcs DY10 1TH
Tel: 01562 754367

Studio Pottery Magazine
15 Magdalen Road
Exeter
Devon EX2 4TA
Tel: 01392 430082

Teddy Bear Times
Ashdown Publishing Ltd
Avalon Court, Star Road
Partridge Green
West Sussex
RH13 8RY

Tel: 01403 711511
Fax: 01403 711521

**The Art & Design
Directory**
Avec Designs Ltd
PO Box 1384
Long Ashton
Bristol
BS18 9DF
Tel: 01275 394639
Fax: 01275 394647

**The Artists & Illustrators
Magazine**
Quarto Magazines Ltd
Level 4
188-194 York Way
London
N7 9QR
Tel: 0171-700 8500
Fax: 0171-700 4985

The Claymaker
Blue Cat House
Publications
14 Brading Road
Brighton
East Sussex BN2 3PD
Tel: 01273 680517

The Craftsman Magazine
PO Box 5
Lowthorpe
Driffield YO25 8JD
Tel: 01377 255213
Fax: 01377 255730
Contact: Angie Boyer

The Glass Engraver
Guild of Glass Engravers
19 Wildwood Road
London
NW11 6UL
Tel: 0181-731 9352

The Quilter
The Quilters Guild
Room 190
Dean Clough
Halifax
West Yorks HX3 5AX
Tel: 01422 347669

The World of Embroidery
The Embroiderers' Guild
PO Box 42B
East Molesey

Surrey
KT8 9BB
Tel: 0181-943 1229
Fax: 0181-977 9882

Toymaking
GMC Publications Ltd
Castle Place
166 High Street
Lewes
East Sussex BN7 1XU
Tel: 01273 477374
Fax: 01273 478606

Traditional Woodworking
Link House Magazines
Dingwall Avenue
Croydon
Surrey CR9 2TA
Tel: 0181-686 2599
Fax: 0181-781 1159

Woodturning
GMC Publications Ltd
Castle Place
166 High Street
Lewes
East Sussex
BN7 1XU
Tel: 01273 477374
Fax: 01273 478606

Woodcarving
GMC Publications Ltd
Castle Place
166 High Street
Lewes
East Sussex
BN7 1XU
Tel: 01273 477374
Fax: 01273 478606

Woodworker
Nexus Special Interests
Nexus House
Boundary Way
Hemel Hempstead
Herts
HP2 7ST
Tel: 01442 66551

Workbox
Workbox Enterprises Ltd
Upcott Hall
Bishops Hull
Taunton
Somerset
TA1 1AQ

INDEX BY CRAFT TYPE

The following is an index constructed by craft type. However certain crafts or names for craft activities are so frequently encountered that they have not been specifically listed.

You will find, therefore, no references to: cards, ceramics, jewellery, paintings, pottery, prints, stoneware or wood as they appear on nearly every page of this work, thus making their specific listing superfluous

I N D E X   O F   P L A C E   N A M E S

CUMBRIA TOURIST BOARD
Ashleigh
Holly Road, Windermere
Cumbria LA23 2AQ
Tel: 015394 44444
covering the county of Cumbria
including the Lake District

EAST OF ENGLAND TOURIST BOARD
Toppesfield Hall
Hadleigh
Suffolk IP7 5DN
Tel: 01473 822922
covering the counties of Bedfordshire,
Cambridgeshire, Essex, Hertfordshire,
Lincolnshire, Norfolk and Suffolk

HEART OF ENGLAND TOURIST BOARD
Woodside, Larkhill Road
Worcester
Worcestershire WR5 2EZ
Tel: 01905 763436
covering the counties of Derbyshire,
Gloucestershire, Hereford and Worces-
ter, Leicestershire, Northamptonshire,
Nottinghamshire, Shropshire, Stafford-
shire, Warwickshire, West Midlands and
West Oxfordshire

NORTH WEST TOURIST BOARD
Swan House
Swan Meadow Road, Wigan Pier
Wigan WN3 5BB
Tel: 01942 821222
covering the counties of Cheshire,
Greater Manchester, Lancashire,
Merseyside and the High Peak District of
Derbyshire

NORTHUMBRIA TOURIST BOARD
Aykley Heads
Durham DH1 5UX
Tel: 0191 375 3000
covering the counites of: Cleveland,
Durham, Northumberland and Tyne &
Wear

**SOUTH EAST ENGLAND TOURIST
BOARD**
The Old Brew House
Warwick Park, Tunbridge Wells
Kent TN2 5TU
Tel: 01892 511008
covering the counties of East and West
Sussex, Kent and Surrey

SOUTHERN TOURIST BOARD
40 Chamberlayne Road
Eastleigh
Hampshire SO50 5JH
Tel: 01703 620006
covering the counties of Berkshire,
Buckinghamshire, Eastern and Northern
Dorset, Hampshire, Oxfordshire and The
Isle of Wight

WEST COUNTRY TOURIST BOARD
60 St. Davids Hill
Exeter
Devon EX4 4SY
Tel: 01392 276351
covering the counties of Devon,
Cornwall, Western Dorset, North and
South Somerset and Wiltshire

YORKSHIRE TOURIST BOARD
312 Tadcaster Road
York YO2 2HF
Tel: 01904 707961
covering the counties of North, South
and West Yorkshire and Humberside

LONDON TOURIST BOARD
26 Grosvenor Gardens
London
SW1W ODU
Tel: 0171 730 3450

NATIONAL TOURIST BOARDS

**NORTHERN IRELAND TOURIST
BOARD**
St Annes Court
59 North Street
Belfast
BT1 1ND
Tel: 01232 231122

SCOTTISH TOURIST BOARD
23 Ravelston Terrace
Edinburgh
EH4 3EU
Tel: 0131 343 1513

WALES TOURIST BOARD
Brunel House
2 Fitzalan Road
Cardiff
CF2 1UY
Tel: 01222 485031

An eye for
Colour Printing...

For inquiries relating to: listings phone **01782 749919**
trade sales or advertising : **0181 304 1164** and retail sales **0181 770 7087**

© **The Write Angle Press 1997**
44 Kingsway, Stoke-on-Trent, Staffordshire ST4 1JH